Hugh Piper

Poultry

A practical guide to the choice, breeding, rearing and management of all descriptions of fowls, turkeys, guinea-fowls, ducks, and geese, for profit and exhibition

Hugh Piper

Poultry

A practical guide to the choice, breeding, rearing and management of all descriptions of fowls, turkeys, guinea-fowls, ducks, and geese, for profit and exhibition

ISBN/EAN: 9783337292010

Printed in Europe, USA, Canada, Australia, Japan

Cover: Foto ©Lupo / pixelio.de

More available books at **www.hansebooks.com**

POULTRY

A

𝔓ractical 𝔊uide

TO THE

CHOICE, BREEDING, REARING, AND MANAGEMENT

OF ALL DESCRIPTIONS OF

FOWLS, TURKEYS, GUINEA-FOWLS, DUCKS, AND GEESE,

FOR

PROFIT AND EXHIBITION.

BY

HUGH PIPER,

AUTHOR OF "PIGEONS: THEIR VARIETIES, MANAGEMENT, BREEDING, AND DISEASES."

ILLUSTRATED WITH EIGHT COLOURED PLATES.

𝔗hird 𝔈dition.

LONDON:
GROOMBRIDGE & SONS, 5, PATERNOSTER ROW,

MDCCCLXXIII.

PREFACE.

This work is intended as a practical guide to those about to commence Poultry keeping, and to provide those who already have experience on the subject with the most trustworthy information compiled from the best authorities of all ages, and the most recent improvements in Poultry Breeding and Management. The Author believes that he has presented his readers with a greater amount of valuable information and practical directions on the various points treated than will be found in most similar works. The book is not the result of the Author's own experience solely, and he acknowledges the assistance he has received from other authorities. Among those whom he has consulted he desires specially to acknowledge his obligations to Mr. Tegetmeier, whose "Poultry Book" (published by Messrs. Routledge & Sons, London) contains his especial knowledge of the Diseases of Poultry; and to Mr. L. Wright, whose excellent and practical Treatise, entitled "The Practical Poultry Keeper" (published by Messrs. Cassell, Petter & Galpin, London), cannot be too highly commended.

CONTENTS.

GENERAL MANAGEMENT.

PAGE

CHAPTER I.—INTRODUCTION 1
Neglect of Poultry-breeding—Profit of Poultry-keeping—Value to the Farmer—Poultry Shows—Cottage Poultry.

CHAPTER II.—THE FOWL-HOUSE 6
Size of the House—Brick and Wood—Cheap Houses—The Roof—Ventilation—Light—Warmth—The Flooring—Perches—Movable Frame—Roosts for Cochin-Chinas and Brahma-Pootras—Nests for laying—Cleanliness—Fowls' Dung—Doors and Entrance-holes—Lime-washing—Fumigating—Raising Chickens under Glass.

CHAPTER III.—THE FOWL-YARD 18
Soil—Situation—Covered Run—Pulverised Earth for deodorising—Diet for confined Fowls—Height of Wall, &c.—Preventing Fowls from flying—The Dust-heap—Material for Shells—Gravel—The Gizzard—The Grass Run.

CHAPTER IV.—FOOD 27
Table of relative constituents and qualities of Food—Barley—Wheat—Oats—Meal—Refuse Corn—Boiling Grain—Indian Corn, or Maize—Buckwheat—Peas, Beans and Tares—Rice—Hempseed—Linseed—Potatoes—Roots—Soft Food—Variety of Food—Quantity—Mode of Feeding—Number of Meals—Grass and Vegetables—Insects—Worms—Snails and Slugs—Animal Food—Water—Fountains.

CONTENTS.

CHAPTER V.—Eggs 40

Eggs all the Year round—Warmth essential to laying—Forcing Eggs—Soft Shells—Shape and Colour of Eggs—The Air-bag—Preserving Eggs—Keeping and Choosing Eggs for setting—Sex of Eggs—Packing Setting-eggs for travelling.

CHAPTER VI.—The Sitting Hen 48

Evil of restraining a Hen from sitting—Checking the Desire—A separate House and Run—Nests for sitting in—Damping Eggs—Filling for Nests—Choosing their own Nests—Choosing a Hen for sitting—Number and Age of Eggs—Food and Exercise—Absence from the Nest—Examining the Eggs—Setting two Hens on the same day—Time of Incubation—The "tapping" sound—Breaking the Shell—Emerging from the Shell—Assisting the Chicken—Artificial Mothers—Artificial Incubation.

CHAPTER VII.—Rearing and Fattening Fowls 63

The Chicken's first Food—Cooping the Brood—Basket and Wooden Coops—Feeding Chickens—Age for Fattening—Barn-door Fattening—Fattening-Houses—Fattening-Coops—Food—"Cramming"—Capons and Poulardes—Killing Poultry—Plucking and packing Fowls—Preserving Feathers.

CHAPTER VIII.—Stock, Breeding, and Crossing 75

Well-bred Fowls—Choice of Breed—Signs of Age—Breeding in-and-in—Number of Hens to one Cock—Choice of a Cock—To prevent Cocks from fighting—Choice of a Hen—Improved Breeds—Origin of Breeds—Crossing—Choice of Breeding Stock—Keeping a Breed pure.

CHAPTER IX.—Poultry Shows 83

The first Show—The first Birmingham Show—Influence of Shows—Exhibition Rules—Hatching for Summer and Winter Shows—Weight—Exhibition Fowls sitting—Matching Fowls—Imparting lustre to the Plumage—Washing Fowls—Hampers—Travelling—Treatment on Return—Washing the Hampers and Linings—Exhibition Points—Technical Terms.

BREEDS.

	PAGE
CHAPTER X.—Cochin-Chinas, or Shanghaes	93
CHAPTER XI.—Brahma-Pootras	101
CHAPTER XII.—Malays	105
CHAPTER XIII.—Game	108
CHAPTER XIV.—Dorkings	112
CHAPTER XV.—Spanish	115
CHAPTER XVI.—Hamburgs	118
CHAPTER XVII.—Polands	121
CHAPTER XVIII.—Bantams	124
CHAPTER XIX.—French and Various	128
CHAPTER XX.—Turkeys	132
CHAPTER XXI.—Guinea-Fowls	139
CHAPTER XXII.—Ducks	142
CHAPTER XXIII.—Geese	147
CHAPTER XXIV.—Diseases	150

LIST OF PLATES.

	PAGE
PLATE I.—Facing the Title-page.	
White Dorking Cock—Coloured Dorkings—Duck-winged and Black-breasted Red Game.	
PLATE II.	93
White and Buff Cochin-China—Malay Cock—Light and Dark Brahma-Pootras.	
PLATE III.	115
Golden-pencilled and Silver-spangled Hamburgs — Black Spanish.	
PLATE IV.	121
White-crested Black Polish — Golden and Silver-spangled Polish.	
PLATE V.	124
White and Black Bantams—Gold and Silver-laced or Sebright Bantams—Game Bantams.	
PLATE VI.	128
French: Houdans—La Flêche Cock—Crêve Cœur Hen.	
PLATE VII.	132
Turkey—Guinea-Fowls.	
PLATE VIII.	142
Toulouse Goose—Rouen Ducks—Aylesbury Ducks.	

PROFITABLE AND ORNAMENTAL
POULTRY.

CHAPTER I.

INTRODUCTION.

UNTIL of late years the breeding of poultry has been almost generally neglected in Great Britain. Any kind of mongrel fowl would do for a farmer's stock, although he fully appreciated the importance of breeding in respect of his cattle and pigs, and the value of improved seeds. Had he thought at all upon the subject, it must have occurred to him that poultry might be improved by breeding from select specimens as much as any other kind of live stock. The French produce a very much greater number of fowls and far finer ones for market than we do. In France, Bonington Mowbray observes, "poultry forms an important part of the live stock of the farmer, and the poultry-yards supply more animal food to the great mass of the community than the butchers' shops"; while in Egypt, and some other countries of the East, from time immemorial, vast numbers of chickens have been hatched in ovens by artificial heat to supply the demand for poultry; but in Great Britain poultry-keeping has been generally neglected, eggs are dear, and all kinds of poultry so great a luxury that the lower classes and a large number of the middle seldom, if ever, taste it, except perhaps once a year in the form of a Christmas goose, while hundreds of thousands cannot afford even this. It is computed that a million of eggs are eaten daily in London and its suburbs alone; yet this vast number only gives one egg to every three months. "It is a national waste," says Mr. Edwards, "importing eggs by the hundreds of millions, and poultry by tens of thousands, when we are feeding our cattle upon corn, and grudging it to our poultry; although the return made from the former, it is generally admitted, is not five per cent. beyond the value of the corn consumed, whereas an immense percentage can be realised by feeding poultry." A writer in the *Times*, of February 1, 1853, states that, while it will take five years to fatten an ox to the weight of sixty stone, which will produce a profit of £30, the same sum may be realised in five months by feeding an equal weight of poultry for the table.

Although fowls are so commonly kept, the proportion to the population is still very small, and the number of those who rear and manage them profitably still smaller, chiefly because most people keep them without system or order, and have not given the slightest attention to the subject. Nevertheless, it costs no more trouble and much less expense to keep fowls successfully and profitably, for neglected fowls are always falling sick, or getting into mischief and causing annoyance, and often expense and loss. "A man," says Mr. Edwards, "who expects a good return of flesh and eggs from fowls insufficiently fed and cared for, is like a miller expecting to get meal from a neglected mill, to which he does not supply grain."

The antiquated idea that fowls on a farm did mischief to the crops has been proved to be false; for if the grain is sown as deeply as it should be, they cannot reach it by scratching; and, besides, they greatly prefer worms and insects. Mr. Mechi says, "commend me to poultry as the farmer's best friend," and considers the value of fowls, in destroying the vast number of worms, grubs, flies, beetles, insects, larvæ, &c., which they devour, as incalculable; and the same may be said as to their destruction of the seeds of weeds. They also consume large quantities of kitchen and table refuse, which is generally otherwise wasted, and often allowed to decay and become a source of disease, or at least of impurity.

The enormous prices paid at the poultry shows of 1852 and 1853 for fancy fowls gave a new impulse to poultry-keeping; and many persons who formerly thought the management of poultry beneath their attention, now superintend their yards. Mrs. Ferguson Blair, now the Hon. Mrs. Arbuthnot, the authoress of the "Henwife," whose experience may be judged by the fact that she gained in four years upwards of 460 prizes in England and Scotland, and personally superintended the management of forty separate yards, in which above 1,000 chickens were hatched annually, says:—

"I began to breed poultry for amusement only, then for exhibition, and lastly, was glad to take the trouble to make it pay, and do not like my poultry-yard less because it is not a loss. It is impossible to imagine any occupation more suited to a lady, living in the country, than that of poultry rearing. If she has any superfluous affection to bestow, let it be on her chicken-kind and it will be returned cent. per cent. Are you a lover of nature? come with me and view, with delighted gaze, her chosen dyes. Are you a utilitarian? rejoice in such an increase of the people's food. Are you a philanthropist? be grateful that yours has been the privilege to afford a *possible* pleasure to the poor man, to whom so many are *impossible.* Such we often find fond of poultry—no mean judges of it, and frequently successful in exhibition. A poor man's pleasure in victory is, at least, as great as that of his richer brother. Let him, then, have the field whereon to fight for it. Encourage village poultry-shows, not only by your patronage, but also by your presence. A taste for such may save many from dissipation and much evil; no man can win poultry honours and haunt the taproom too."

For those who desire to encourage a taste for poultry keeping in young people, and their humbler neighbours, we would recommend our smaller work on the subject as a suitable present.*

---

* Piper on Poultry : their Varieties, Management, Breeding, and Diseases. Price 1s. Groombridge & Sons, 5, Paternoster Row, London.

"It becomes," says Miss Harriet Martineau, "an interesting wonder every year why the rural cottagers of the United Kingdom do not rear fowls almost universally, seeing how little the cost would be and how great the demand. We import many millions of eggs annually. Why should we import any? Wherever there is a cottage family living on potatoes or better fare, and grass growing anywhere near them, it would be worth while to nail up a little penthouse, and make nests of clean straw, and go in for a speculation in eggs and chickens. Seeds, worms, and insects go a great way in feeding poultry in such places; and then there are the small and refuse potatoes from the heap, and the outside cabbage leaves, and the scraps of all sorts. Very small purchases of broken rice (which is extremely cheap), inferior grain, and mixed meal, would do all else that is necessary. There would be probably larger losses from vermin than in better guarded places; but these could be well afforded as a mere deduction from considerable gains. It is understood that the keeping of poultry is largely on the increase in the country generally, and even among cottagers; but the prevailing idea is of competition as to races and specimens for the poultry-yard, rather than of meeting the demand for eggs and fowls for the table."

With the exception of prizes for Dorkings, which are chiefly bred for market, our poultry-shows have always looked upon fowls as if they were merely ornamental birds, and have framed their standards of excellence accordingly, and not with any regard to the production of profitable poultry, which is much to be regretted.

Martin Doyle, the cottage economist of Ireland, in his "Hints to Small Holders," observes that "a few cocks and hens, if they be prevented from scratching in the garden, are a useful and appropriate stock about a cottage, the warmth of which causes them to lay eggs in winter—no trifling advantage to the children when milk is scarce. The French, who are extremely fond of eggs, and contrive to have them in great abundance, feed the fowls so well on curds and buckwheat, and keep them so warm, that they have plenty of eggs even in winter. Now, in our country

(Ireland), especially in a gentleman's fowl yard, there is not an egg to be had in cold weather; but the warmth of the poor man's cabin insures him an egg even in the most ungenial season."

Such fowls obtain fresh air, fresh grass, and fresh ground to scratch in, and prosper in spite of the most miserable, puny, mongrel stock, deteriorating year after year from breeding in and in, without the introduction of fresh blood even of the same indifferent description. Many an honest cottager might keep himself and family from the parish by the aid of a small stock of poultry, if some kind poultry-keeper would present him with two or three good fowls to begin with, for the cottager has seldom capital even for so small a purchase.

Considerable profit may be made by the sale of eggs for hatching and surplus stock, if the breeds kept are good, and the stock known to be pure and vigorous. The "Henwife" says: "You may reduce your expenses by selling eggs for setting, at a remunerative price. No one should be ashamed to own what he is not ashamed to do; therefore, boldly announce your superfluous eggs for sale, at such a price as you think the public will pay for them." This is now done extensively by breeders of rank and eminence, especially through the London *Field* and agricultural papers. But, "beware of sending such eggs to market. Every one would be set, and you might find yourself beaten by your own stock, very likely in your own local show, and at small cost to the exhibitor."

The great secret of success in keeping fowls profitably is to hatch chiefly in March and April; encourage the pullets by proper feeding to lay at the age of six months; and fatten and dispose of them when about nineteen months old, just before their first adult moult; and never to allow a cockerel to exceed the age of fourteen weeks before it is fattened and disposed of.

CHAPTER II.

THE FOWL-HOUSE.

In this work we shall consider the accommodation and requisites for keeping fowls successfully on a moderate scale, and the reader must adapt them to his own premises, circumstances, and requirements. Everywhere there must be some alterations, omissions, or compromises. We shall state the essentials for their proper accommodation, and describe the mode of constructing houses, sheds, and arranging runs, and the reader must then form his plan according to his own wishes, resources, and the capabilities of the place. The climate of Great Britain being so very variable in itself, and differing in its temperature so much in different parts, no one manner or material for building the fowl-house can be recommended for all cases.

Plans for poultry establishments on large scales for the hatching, rearing, and fattening of fowls, turkeys, ducks, and geese, are given in our smaller work on Poultry, referred to on page 3.

The best aspects for the fowl-house are south and south-east, and sloping ground is preferable to flat.

"It is only of late years," says Mr. Baily, "poultry-houses have been much thought of. In large farmyards, where there are cart-houses, calf-pens, pig-styes, cattle-sheds, shelter under the eaves of barns, and numerous other roosting-places, not omitting the trees in the immediate vicinity, they are little required—fowls will generally do better by choosing for themselves; and it is beyond a doubt healthier for them to be spread about in this manner, than to be confined to one place. But a love of order, on the one hand, and a dread of thieves or foxes on the other, will sometimes make it desirable to have a proper poultry-house."

Size of the House.

Each family of fowls should, if possible, have a house and run; and if they are kept as breeding stock, and the breeds are to be preserved pure, this is essential. And where many kinds are kept, the various houses must be adapted to the peculiarities of the different breeds, in order to do justice to them all, and to attain success in each.

The size of the house and the extent of the yard or run should be proportioned to the number of fowls kept; but it is better for the house to be too small than too large, particularly in winter, for the mutual imparting of animal heat. It is found by experience that when fowls are crowded into a small space, their desire for laying continues even in winter; and there is no fear of engendering disease by crowding if the house is properly ventilated, and thoroughly cleansed every day. Mr. Baily kept for years a cock and four hens in a portable wooden house six feet square, and six feet high in the centre, the sides being somewhat shorter, and says such a house would hold six hens as well as four. Ventilating holes were made near the top. It had no floor, being placed upon the ground, and could be moved at pleasure by means of two poles placed through two staples fixed at the end of each side. A few Cochin-Chinas may be kept where there is no other convenience than an outhouse six feet square to serve for their roosting, laying, and sitting, with a yard of twice that size attached. Mr. Wright "once knew a young man who kept fowls most profitably, with only a house of his own construction, not more than three feet square, and a run of the same width, under twelve feet long." The French breeders keep their fowls in as small a space as possible, in order to generate and preserve the warmth that will induce them to lay; while the English breeders allow more space for exercise, larger houses, and free circulation of air. The French mode is very likely the best for the winter and the English for the summer, but the two opposite methods may be made available by having one or more extra houses and runs into which the fowls can be distributed in the summer. A close, warm roosting-place will cause the production of more eggs in winter, when they are scarcest and most valuable, while air and exercise are necessary to rear superior fowls for the table; and if they can have the run of a farmyard or good fields in which to pick up grain or insects, their flesh will be far superior in flavour to that of fowls kept in confinement, or crammed in coops.

Almost any outbuilding, shed, or lean-to, may be easily and cheaply converted into a good fowl-house by the exercise of a little thought and ingenuity.

The best material to build a house with is brick, but the cheapest to be durable is board, with the roof also of wood, covered with patent felt. One objection to timber houses is their being combustible, and easily ignited, and houses had better be built of a single brick in thickness, unless cheapness is a great object.

A lean-to fowl-house may be constructed for a very small sum, with boards an inch thick, against the west or south side of any wall. Whenever wood is employed it should be tongued, which is a very cheap method of providing against warping by heat, or admitting wind or rain; lying flat against the uprights, it saves material and has an external appearance far superior to any other method of boarding. If the second coat of paint is rough cast over with sand, it will greatly improve the appearance, and the house will not be unsightly even in the ornamental part of a gentleman's grounds.

A house may be built very cheaply by driving poles into the ground at equal distances, and nailing weather-boarding upon their outside. If it is to be square, one pole should be placed at each corner, and two more will be required for the door-posts. The house may be made with five, six, or more sides, as many poles being used as there are sides, and the door may occupy one side if the house be small and the side narrow, otherwise two door-posts will be required. If the boards are not tongued together, the chinks between them must be well caulked by driving in string or tow with a blunt chisel, for it is not only necessary to keep out the rain but also to keep out the wind, which has great influence on the health and laying of the fowls.

Where double boarding is employed for the sides, the house may be made much warmer by filling up the space with straw, or still better with marsh reeds, so durable for thatching. This plan, unfortunately, affords a shelter for rats, mice, and insects, and therefore, if adopted, it will be highly advantageous to form the inside boarding in panels, so as to be removable at pleasure for examination and cleansing.

For the roof, tiles or slates alone are not sufficient, but, if used, must have a boarding or ceiling under them; otherwise all the heat generated by the fowls will escape through the numerous interstices, and it will be next to impossible to keep the house warm in winter. A corrugated roof of galvanised iron may be used instead, but a ceiling also will be absolutely necessary for the sake of warmth. A rough ceiling of lath and plaster not only preserves the warmth generated by the fowls and keeps out the cold, but has the great advantage of being easily lime-washed, an operation that should be performed at least four or five times a year. Boards alone make a very good and cheap roof. They may be laid either horizontally, one plank overlapping the other, and the whole well tarred two or three times, and once every autumn afterwards; or they may be laid perpendicularly side by side, fitting closely, in which case they should be well tarred, then covered with old sheeting, waste calico, or thick brown paper tightly stretched over it, and afterwards brushed over with hot tar, or a mixture of tar boiled with a little lime, and applied while hot; this, soaking through the calico, cements it to the roof, and makes it waterproof. But board covered with patent felt, and tarred once a year, is the best. The roof ought to project considerably beyond the walls, in order to prevent the rain from dripping down them.

Ventilation is most important, and the house should be high, especially if there are many fowls, for by having it lofty a current of air can pass through it far above the level of the fowls, and purify the atmosphere without causing a draught near them. They very much dislike a draught, and will alter their positions to avoid it, and if unable to do so, will seek another roosting-place. Ventilation may be obtained by leaving out some bricks in the wall or making holes in the boarding; and when there is a shed at the side of the fowl-house, by boring a few holes near the top of the wall next to the shed; all ventilators should be considerably above the perches, in order to avoid a draught near to the fowls; and should be entirely closed at night in severe weather. The best method of ventilation for a fowl-house of sufficient size and height, is by means of an opening in the highest part of the roof, covered with a lantern of laths or narrow boards, placed one over the other in a slanting position, with a small space between them like Venetian blinds.

Light is essential, not only for the health of the fowls, but in order that the state of the house may be seen, and the floor and perches may be well cleansed. It may be admitted either through a common window, a pane or two of thick glass placed in the sides, or glass tiles in the roof. It also induces them to take shelter there in rough weather.

Warmth is the most important point of all. Fowls that roost in cold houses and exposed places require more food and produce fewer eggs; and pullets which are usually forward in laying will not easily be induced to do so in severe weather if their house is not kept warm. It is a great advantage when the house backs a fire-place or stable. A gentleman told Mr. Baily that he "had been very successful in raising early chickens in the north of Scotland, and he attributed much of it to the following arrangements. He had always from twenty to thirty oxen or other cattle fattening in a long building; he made his poultry-house to join this, and had ventilators and openings made in the partition, so that the heat of the cattle-shed passed into the fowl-house. Little good has resulted from the use of stoves, or hot-water pipes, for poultry; but by skilfully taking advantage of every circumstance like that above mentioned, and by consulting aspect and position, many valuable helps are obtained."

A house built of wood in the north of England and

Scotland must be lined, unless artificially warmed. Felt is the best material, as its strong smell of tar will keep away most insects. Matting is frequently used, and will make the house sufficiently warm, but it harbours vermin, and therefore, if used, should be only slightly fastened to the walls, so that it can be often taken down and well beaten, and, if necessary, fumigated.

Various materials are recommended for the flooring. Boards are warm, but they soon become foul. Beaten earth, with loose dust scattered over it some inches deep, is excellent for the feet of the birds, but is a harbour for the minute vermin which are often so troublesome, and even destructive, to domestic fowls. Mowbray recommends a floor of "well-rammed chalk or earth, that its surface, being smooth, may present no impediment to being swept perfectly clean." Chalk laid on dry coal-ashes to absorb the moisture is excellent. A mixture of cow-dung and water, about the consistency of paint, put on the surface of the floor, no thicker than paint, gives it a hard surface which will bear sweeping down. It is used by the natives of India, not only for the floors, but often for the walls of their houses, and is supposed to be healthy in its application, and to keep away vermin. Miss Watts says: "Dig out the floor to about a foot deep, and fill in with burnt clay, like that used extensively on railways, the strong gravel which is called 'metal' in road-making, or any loose dry material of the kind. Let this be well rammed down, and then lay over it, with a bricklayer's trowel, a flooring of a compost of cinder-ashes, gravel, quick-lime, and water. This flooring is without the objections due to those which are cold and damp, and those which imbibe foul moisture. Stone is too cold for a flooring; beaten earth or wood becomes foul when the place is inhabited by living animals; and a flooring of bricks possesses both these bad qualities united." Bricks are the worst of all materials; they retain moisture, whether atmospheric or arising from insufficient drainage; and thus the temperature is kept low, and disease too often follows, especially rheumatic attacks of the feet and legs. However, trodden earth makes a very good flooring, and it or other materials may easily be kept clean by placing moveable boards beneath the perches to receive the fowl-droppings. The floor should slope from every direction towards the door, to facilitate its cleansing, and to keep it dry.

Perches are generally placed too high, probably because it was noticed that fowls in their natural state, or when at large, usually roost upon high branches; but it should be observed that, in descending from lofty branches, they have a considerable distance to fly, and therefore alight on the ground gently, while in a confined fowl-house the bird flutters down almost perpendicularly, coming into contact with the floor forcibly, by which the keel of the breast-bone is often broken, and bumble-foot in Dorkings and corns are caused.

Some writers do not object to lofty perches, provided the fowls have a board with cross-pieces of wood fastened on to it reaching from the ground to the perch; but this does not obviate the evil, for they will only use it for ascent, and not for descent. The air, too, at the upper part of any dwelling-room, or house for animals, is much more impure than nearer the floor, because the air that has been breathed, and vapours from the body, are lighter than pure air, and consequently ascend to the top. The perches should therefore not be more than eighteen inches from the ground, unless the breed is very small and light. Perches are also generally made too small and round. When they are too small in proportion to the size of the birds, they are apt to cause the breast-bone of heavy fowls to grow crooked, which is a great defect, and very unsightly in a table-fowl. Those for heavy fowls should not be less than three inches in diameter. Capital perches may be formed of fir or larch poles, about three inches in diameter, split into two, the round side being placed uppermost; the birds' claws cling to it easily, and the bark is not so hard as planed wood. The perches, if made of timber, should be nearly square, with only the corners rounded off, as the feet of fowls are not formed for clasping smooth round poles. Those for chickens should not be thicker than their claws can easily grasp, and neither too sharp nor too round.

When more than one row of perches is required they should be ranged obliquely—that is, one above and behind the other; by which arrangement each perch forms a step to the next higher one, and an equal convenience in descending, and the birds do not void their dung over each other. They should be placed two feet apart, and supported on bars of wood fixed to the walls at each end; and in order that they may be taken out to be cleaned, they should not be nailed to the supporter, but securely placed in niches cut in the bar, or by pieces of wood nailed to it like the rowlocks of a boat. If the wall space at the sides is required for laying-boxes, the perches must be shorter than the house, and the oblique bars which support them must be securely fastened to the back of the house, and, if necessary, have an upright placed beneath the upper end of each.

Some breeders prefer a moveable frame for roosting, formed of two poles of the required length, joined at each end by two narrow pieces; the frame being supported upon four or more legs, according to its length and the weight of the fowls. If necessary it should be strengthened by rails—connecting the bottoms of the legs, and by pieces crossing from each angle of the sides and ends. These frames can conveniently be moved out of the house when they require cleansing. Or it may be made of one pole supported at each end by two legs spread out widely apart, like two sides of an equilateral or equal-sided triangle. The perch may be made more secure for heavy fowls by a rail at each side fastened to each leg, about three inches from the foot.

Mr. Baily says: "I had some fowls in a large outhouse, where they were well provided with perches; as there was plenty of room, I put some small faggots, cut for firing, at one extremity, and I found many of the fowls deserted their perches to roost on the faggots, which they evidently preferred."

Cochin-Chinas and Brahma Pootras do not require

perches, but roost comfortably on a floor littered down warmly with straw. It should be gathered up every morning, and the floor cleaned and kept uncovered till night, when the straw, if clean, should be again laid down. It must be often changed. A bed of sand is also used, and a latticed floor even without straw, and some use latticed benches raised about six inches from the floor. But we should think that latticed roosting-places must be uncomfortable to fowls, and the dung which falls through is often unseen, and, consequently, liable to remain for too long a time, while a portion will stick to the sides of the lattice-work, and be not only difficult to see, but also to remove when seen. The "Henwife" finds, however, "that if there are nests, there the Cochins will roost, in spite of all attempts to make them do otherwise." It is a good plan, in warm weather, occasionally to sprinkle water over and about the perches, and scatter a little powdered sulphur over the wetted parts, which will greatly tend to keep the fowls free from insect parasites.

The nests for laying in are usually made on the ground, or in a kind of trough, a little raised; but some use boxes or wicker-baskets, which are preferable, as they can be removed separately from time to time, and thoroughly cleansed from dust and vermin, and can also be kept a little apart from each other. These boxes or troughs should be placed against the sides of the house, and a board sloping forwards should be fixed above, to prevent the fowls from roosting upon the edges. If required, a row of laying-boxes or troughs may be placed on the ground, and another about a foot or eighteen inches above the floor. The nest should be made of wheaten, rye, or oaten straw, but never of hay, which is too hot, and favourable besides to the increase of vermin. Heath cut into short pieces forms excellent material for nests, but it cannot always be had. The material must be changed whenever it smells foul or musty, for if it is allowed to become offensive, the hens will often drop their eggs upon the ground sooner than go to the nest. When the fowl-house adjoins a passage, or it can be otherwise so contrived, it is an excellent plan to have a wooden flap made to open just above the back of the nests, so that the eggs can be removed without your going into the roosting-house, treading the dung about, and disturbing any birds that may be there, or about to enter to lay. Where possible the nests in the roosting-houses should be used for laying in only; and a separate house should be set apart for sitting hens. Where there are but a few fowls and only one house, if a hen is allowed to sit, a separate nest must be made as quiet as possible for her.—*See* Chapter VI.

Cleanliness must be maintained. The *Canada Farmer* suggested an admirable plan for keeping the roosting-house clean. A broad shelf, securely fastened, but moveable, is fixed at the back of the house, eighteen inches from the ground, and the perch placed four or five inches above it, a foot from the wall. The nests are placed on the ground beneath the board, which preserves them from the roosting fowl's droppings, and keeps them well shaded for the laying or sitting hen, if the latter is obliged to incubate in the same house, and the nests do not need a top. The shelf can be easily scraped clean every morning, and should be lightly sanded afterwards. Thus the floor of the house is never soiled by the roosting birds, and the broad board at the same time protects them from upward draughts of air. Where the nests and perches are not so arranged, the idea may be followed by placing a loose board below each perch, upon which the dung will fall, and the board can be taken up every morning and the dung removed. With proper tools, a properly constructed fowl-house can be kept perfectly clean, and all the details of management well carried out without scarcely soiling your hands. A birch broom is the best implement with which to clean the house if the floor is as hard as it ought to be. A handful of ashes or sand, sprinkled over the places from which dung has been removed, will absorb any remaining impurity.

Fowls' dung is a very valuable manure, being strong, stimulating, and nitrogenous, possessing great power in forcing the growth of vegetables, particularly those of the cabbage tribe, and is excellent for growing strawberries, or indeed almost any plants, if sufficiently diluted; for, being very strong, it should always be mixed with earth. A fowl, according to Stevens, will void at least one ounce of dry dung in twenty-four hours, which is worth at least seven shillings a cwt.

The door should fit closely, a slight space only being left at the bottom to admit air. It should have a square hole, which is usually placed either at the top or bottom, for the poultry to enter to roost. A hole at the top is generally preferred, as it is inaccessible to vermin. The fowls ascend by means of a ladder formed of a slanting board, with strips of wood nailed across to assist their feet; a similar ladder should be placed inside to enable them to descend, if they are heavy fowls; but the evil is that, even with this precaution, they are inclined to fly down, as they do from high perches, without using the ladder, and thus injure their feet. A hole in the middle of the door would be preferable to either, and obviate the defects of both. These holes should be fitted with sliding panels on the inside, so that they can be closed in order to keep the fowls out while cleaning the house, or to keep them in until they have laid their eggs, or it may be safe to let them out in the morning in any neighbourhood or place where they would else be liable to be stolen. Every day, after the fowls have left their roosts, the doors and windows should be opened, and a thorough draught created to purify the house. During the winter months all the entrance holes should be closed from sunset to sunrise, unless in mild localities. Where there are many houses, they should, if possible, communicate with each other by doors, so that they may be cleaned from end to end, or inspected without the necessity of passing through the yards, which is especially unpleasant in wet weather. The doors should be capable of being fastened on either side, to avoid the chance of the different breeds intermingling while your attention is occupied in arranging the nests, collecting eggs, &c. See that your fowls are securely locked in at night, for they are more easily stolen than any other kind of domestic animals.

Destroying Vermin.

A good dog in the yard or adjoining house or stable is an excellent protection.

Every poultry-house should be lime-washed at least four or five times a year, and oftener if convenient. Vermin of any kind can be effectually destroyed by fumigating the place with sulphur. In this operation a little care is requisite; it should be commenced early in the morning, by first closing the lattices, and stopping up every crevice through which air can enter; then place on the ground a pan of lighted charcoal, and throw on it some brimstone broken into small pieces. Directly this is done the room should be left, the door kept shut and airtight for some hours; care too should be taken that the lattices are first opened, and time given for the vapour to thoroughly disperse before any one again enters, when every creature within the building will be found destroyed.

It is said that a pair of caged guineapigs in the fowlhouse will keep away rats.

In a large establishment, and in a moderate one, if the outlay is not an object, the pens for the chickens and the passages between the various houses may be profitably covered with glass, and grapes grown on the rafters. Raising chickens under glass has been tried with great success.

CHAPTER III.

THE FOWL-YARD.

THE scarcity of poultry in this country partly arises from all gallinaceous birds requiring warmth and dryness to keep them in perfect health, while the climate of Great Britain is naturally moist and cold.

"The warmest and driest soils," says Mowbray, "are the best adapted to the breeding and rearing of gallinaceous fowls, more particularly chickens. A wet soil is the worst, since, however ill affected fowls are by cold, they endure it better than moisture. Land proper for sheep is generally also adapted to the successful keeping of poultry and rabbits."

But poultry may be reared and kept successfully even on bad soils with good drainage and attention. The "Henwife" says: "I do not consider any one soil necessary for success in rearing poultry. Some think a chalk soil essential for Dorkings, but I have proved the fallacy of this opinion by bringing up, during three years, many hundreds of these *soi disant* delicate birds on the strong blue clay of the Carse of Gowrie, doubtless thoroughly drained, that system being well understood and universally practised by the farmers of the district. A coating of gravel and sand once a year is all that is requisite to secure the necessary dryness in the runs." The best soil for a poultry-yard is gravel, or sand resting on chalk or gravel. When the soil is clayey, or damp from any other cause, it should be thoroughly drained, and the whole or a good portion of the ground should be raised by the addition of twelve inches of chalk, or bricklayer's rubbish, over which should be spread a few inches of sand. Cramp, roup, and some other diseases, more frequently arise from stagnant wet in the soil than from any other cause.

Sheds.

The yard should be sheltered from the north and east winds, and where this is effected by the position of a shrubbery or plantation in which the fowls may be allowed to run, it will afford the advantage of protection, not only from wind and cold, but also shelter from the rain and the burning sun. It also furnishes harbourage for insects, which will find them both food and exercise in picking up. Indeed, for all these purposes a few bushes may be advantageously planted in or adjoining any poultry-yard. When a tree can be enclosed in a run, it forms an agreeable object for the eye, and affords shelter to the fowls.

A covered run or shed for shelter in wet or hot weather is a great advantage, especially if chickens are reared. It may be constructed with a few rough poles supporting a roof of patent felt, thatch, or rough board, plain or painted for preservation, and may be made of any length and width, from four feet upwards, and of any height from four feet at the back and three feet in the front, to eight feet at the back and six feet in the front. The shed should, if possible, adjoin the fowl-house. It should be wholly or partly enclosed with wirework, which should be boarded for a foot from the ground to keep out the wet and snow, and to keep in small chickens. The roof should project a foot beyond the uprights which support it, in order to throw the rain well off, and have a gutter-shoot to carry it away and prevent it from being blown in upon the enclosed space. The floor should be a little higher than the level of the yard, both in order to keep it dry and the easier to keep it clean; and it should be higher at the back than in the front, which will keep it drained if any wet should be blown in or water upset. If preferred, moveable netting may be used, so that the fowls can be allowed their liberty in fine weather, and be confined in wet weather. But the boarding must be retained to keep out the wet. The ground may be left in its natural state for the fowls to scratch in, in which case the surface should be dug up from time to time and replaced with fresh earth pressed down moderately hard. If the house is large and has a good window, a shed is not absolutely necessary, especially for a few fowls only, but it is a valuable addition, and is also very useful to shelter the coops of the mother hens and their young birds in wet, windy, or hot weather.

By daily attention to cleanliness, a few fowls may be kept in such a covered shed, without having any open run, by employing a thick layer of dry pulverised earth as a deodoriser, which is to be turned over with a rake every day, and replaced with fresh dry pulverised earth once a week. The dry earth entirely absorbs all odour. In a run of this kind, six square feet should be allowed to each fowl kept, for a smaller surface of the dry earth becomes moist and will then no longer deodorise the dung. Sifted ashes spread an inch deep over the floor of the whole shed will be a good substitute if the dry earth cannot be had. They should be raked over every other morning, and renewed at least every fortnight, or oftener if possible. The ground should be dug and turned over whenever it looks sodden, or gives out any offensive smell; and three or four times a year the polluted soil below the layer, that is, the earth to the depth of three or four inches, should be removed and replaced with fresh earth, gravel, chalk, or ashes.* The shed must be so contrived that the sun can shine upon the fowls during some part of the day, or they will not continue in health for any length of time, and it is almost impossible to rear healthy chickens without its light and warmth; and it will be a great improvement if part of the run is open. Another shed will be required if chickens are to be reared.

Fowls that are kept in small spaces or under covered runs will require a different diet to those that are allowed to roam in fields and pick up insects, grass, &c., and must be provided with green food, animal food in place of insects, and be well supplied with mortar rubbish and gravel.

The height of the wall, paling, or fencing that surrounds the yard, and of the partitions, if the yard is divided into compartments for the purpose of keeping two or more breeds separate and pure, must be according to the nature of the breed. Three feet in height will be sufficient to retain Cochins and Brahmas; six feet will be required for moderate-sized fowls; and eight or nine feet will be neces-

*The Practical Poultry Keeper. By Mr. L. Wright. Cassell, Petter & Galpin.

sary to confine the Game, Hamburg, and Bantam breeds. Galvanised iron wire-netting is the best material, as it does not rust, and will not need painting for a long time. It is made of various degrees of strength, and in different forms, and may be had with meshes varying from three-fourths of an inch to two inches or more; with very small meshes at the lower part only, to keep out rats and to keep in chickens; with spikes upon the top, or with scolloped wire-work, which gives it a neat and finished appearance; with doors, and with iron standards terminating in double spikes to fix in the ground, by which wooden posts are divided, while it can be easily fixed and removed. The meshes should not be more than two inches wide, and if the meshes of the lower part are not very small, it should be boarded to about two feet six inches from the ground, in order to keep out rats, keep in chickens, and to prevent the cocks fighting through the wire, which fighting is more dangerous than in the open, for the birds are very liable to injure themselves in the meshes, and, Dorkings especially, to tear their combs and toes in them. If iron standards are not attached to the netting, it should be stretched to stout posts, well fixed in the ground, eight feet apart, and fastened by galvanised iron staples. A rail at the top gives a neater appearance, but induces the fowls to perch upon it, which may tempt them to fly over.

Where it is not convenient to fix a fence sufficiently high, or when a hen just out with her brood has to be kept in, a fowl may be prevented from flying over fences by stripping off the vanes or side shoots from the first-flight feathers of one wing, usually ten in number, which will effectually prevent the bird from flying, and will not be unsightly, as the primary quills are always tucked under the others when not used for flying. This method answers much better than clipping the quills of each wing, as the cut points are liable to inflict injuries and cause irritation in moulting.

The openness of the feathers of fowls which do not throw off the water well, like those of most birds, enables them to cleanse themselves easier from insects and dirt, by dusting their feathers, and then shaking off the dirt and these minute pests with the dust. For this purpose one or more ample heaps of sifted ashes, or very dry sand or earth, for them to roll in, must be placed in the sun, and, if possible, under shelter, so as to be warm and perfectly dry. Wood ashes are the best. This dust-heap is as necessary to fowls as water for washing is to human beings. It cleanses their feathers and skin from vermin and impurities, promotes the cuticular or skin excretion, and is materially instrumental in preserving their health. If they should be much troubled with insects, mix in the heap plenty of wood ashes and a little flour of sulphur.

A good supply of old mortar-rubbish, or similar substance, must be kept under the shed, or in a dry place, to provide material for the eggshells, or the hens will be liable to lay soft-shelled eggs. Burnt oyster-shells are an excellent substitute for common lime, and should be prepared for use by being heated red-hot, and when cold broken into small pieces with the fingers, but not powdered. Some give chopped or ground bones, or a lump of chalky marl. Eggshells roughly crushed are also good, and are greedily devoured by the hens.

A good supply of gravel is also essential, the small stones which the fowls swallow being necessary to enable them to digest their hard food. Fowls swallow all grain whole, their bills not being adapted for crushing it like the teeth of the rabbit or the horse, and it is prepared for digestion by the action of a strong and muscular gizzard, lined with a tough leathery membrane, which forms a remarkable peculiarity in the internal structure of fowls and turkeys. "By the action," says Mr. W. H. L. Martin, "of the two thick muscular sides of this gizzard on each other, the seeds and grains swallowed (and previously macerated in the crop, and there softened by a peculiar secretion oozing from glandular pores) are ground up, or triturated in order that their due digestion may take place. It is a remarkable fact that these birds are in the habit of swallowing small pebbles, bits of gravel, and similar substances, which it would seem are essential to their health. The definite use of these substances, which are certainly ground down by the mill-like action of the gizzard, has been a matter of difference among various physiologists, and many experiments, with a view to elucidate the subject, have been undertaken. It was sufficiently proved by Spallanzani that the digestive fluid was incapable of dissolving grains of barley, &c., in their unbruised state; and this he ascertained by filling small hollow and perforated balls and tubes of metal or glass with grain, and causing them to be swallowed by turkeys and other fowls; when examined, after twenty-four and forty-eight hours, the grains were found to be unaffected by the gastric fluid; but when he filled similar balls and tubes with bruised grains, and caused them to be swallowed, he found, after a lapse of the same number of hours, that they were more or less dissolved by the action of the gastric juice. In other experiments, he found that metallic tubes introduced into the gizzard of common fowls and turkeys, were bruised, crushed, and distorted, and even that sharp-cutting instruments were broken up into blunt fragments without having produced the slightest injury to the gizzard. But these experiments go rather to prove the extraordinary force and grinding powers of the gizzard, than to throw light upon the positive use of the pebbles swallowed; which, after all, Spallanzani thought were swallowed without any definite object, but from mere stupidity. Blumenbach and Dr. Bostock aver that fowls, however well supplied with food, grow lean without them, and to this we can bear our own testimony. Yet the question, what is their precise effect? remains to be answered. Boerhave thought it probable that they might act as absorbents to superabundant acid; others have regarded them as irritants or stimulants to digestion; and Borelli supposed that they might really contribute some degree of nutriment."

Sir Everard Home, in his "Comparative Anatomy," says: "When the external form of this organ is first attentively examined, viewing that side which is anterior in the living bird, and on which the two bellies of the muscle and middle are more distinct, there being no other part to obstruct the view, the belly of the muscle on the left side is seen to be larger than on the right. This appears, on reflection, to be of great advantage in producing the necessary motion; for if the two muscles were of equal strength, they must keep a greater degree of exertion than is necessary; while, in the present case, the principal effect is produced by that of the left side, and a smaller force is used by that on the right to bring the parts back again. The two bellies of the muscle, by their alternate action, produce two effects—the one a constant friction on the contents of the cavity; the other, a pressure on them. This last arises from a swelling of the muscle inwards, which readily explains all the instances which have been given by Spallanzani and others, of the force of the gizzard upon substances introduced into it—a force which is found by their experiments always to act in an oblique direction. The internal cavity, when opened in this distended state, is found to be of an oval form, the long diameter being in the line of the body; its capacity nearly equal to the size of a pullet's egg; and on the sides there are ridges in their horny coat (lining membrane) in the long direction of the oval. When the horny coat is examined in its internal structure, the fibres of which it is formed are not found in a direction perpendicular to the ligamentous substance behind it; but in the upper portion of the cavity it is obliquely upwards. From this form of cavity it is evident that no part of the sides is ever intended to be brought in contact, and that the food is triturated by being mixed with hard bodies, and acted on by the powerful muscles which form the gizzard."

The experiments of Spallanzani show that the muscular action of the gizzard is equally powerful whether the small stones are present or not; and that they are not at all necessary to the trituration of the firmest food, or the hardest foreign substances; but it is also quite clear that when these small stones are put in motion by the muscles of the gizzard they assist in crushing the grain, and at the same time prevent it from consolidating into a thick, heavy, compacted mass, which would take a far longer time in undergoing the digestive process than when separated and intermingled with the pebbles.

This was the opinion of the great physiologist, John Hunter, who, in his treatise "On the Animal Economy," after noticing the grinding powers of the gizzard, says, in reference to the pebbles swallowed, "We are not, however, to conclude that stones are entirely useless; for if we compare the strength of the muscles of the jaws of animals which masticate their food with those of birds who do not, we shall say that the parts are well calculated for the purpose of mastication; yet we are not thence to infer that the teeth in such jaws are useless, even although we have proof that the gums do the business when the teeth are gone. If pebbles are of use, which we may reasonably conclude they are, birds have an advantage over animals having teeth, so far as pebbles are always to be found, while the teeth are not renewed. If we constantly find in an organ substances which can only be subservient to the functions of that organ, should we deny their use, although the part can do its office without them? The stones assist in grinding down the grain, and, by separating its parts, allow the gastric juice to come more readily in contact with it."

When a paddock is used as a run for a large number of poultry, it should be enclosed either by a wall or paling, but not by a hedge, as the fowls can get through it, and will also lay their eggs under the hedge. The paddock should be well drained, and it will be a great advantage if it contains a pond, or has a stream of water running through or by it. Mowbray advises that the grass run should be sown "with common trefoil or wild clover, with a mixture of burnet, spurry, or storgrass," which last two kinds "are particularly salubrious to poultry." If the grass is well rooted before the fowls are allowed to run on it, they may range there for several hours daily, according to its extent and their number, but it should be renewed in the spring by sowing where it has become bare or thin. A dry common, or pasture fields, in which they may freely wander and pick up grubs, insects, ants' eggs, worms, and leaves of plants, is a great advantage, and they may be accustomed to return from it at a call. Where there is a cropped field, orchard, or garden, in which fowls may roam at certain seasons, when the crops are safe from injury, each brood should be allowed to wander in it separately for a few hours daily, or on different days, as may be most convenient. "A garden dung-heap," says Mr. Baily, "overgrown with artichokes, mallows, &c., is an excellent covert for chickens, especially in hot weather. They find shelter and meet with many insects there." When horse-dung is procured for the garden, or supplied from your stables, some should be placed in a small trench, and frequently renewed, in which the fowls will amuse themselves, particularly in winter, by scraping for corn and worms. When fowls have not the advantage of a grass run they should be indulged with a square or two of fresh turf, as often as it can be obtained, on which they will feed and amuse themselves. It should be heavy enough to enable them to tear off the grass, without being obliged to drag the turf about with them.

CHAPTER IV.

FOOD.

The following table, which first appeared ι the " Poultry Diary," will show at a glance the relative constituents and qualities of the different kinds of food, and may be consulted with great advantage by the poultry-keeper, as it will enable him to proportion mixed food correctly, and to change it according to the production of growth, flesh, or fat that may be desired, and according to the temperature of the season. These proportions, of course, are not absolutely invariable, for the relative proportions of the constituents of the grain will vary with the soil, manure used, and the growing and ripening characteristics of the season.

There is in every 100 lbs. of	Flesh-forming Food.	Warmth-giving Food.		Bone-making Food.	Husk or Fibre.	Water.
	Gluten, &c.	Fat or Oil.	Starch, &c.	Mineral Substance		
Oats	15	6	47	2	20	10
Oatmeal	18	6	63	2	2	9
Middlings or fine Sharps	18	6	53	5	4	14
Wheat	12	3	70	2	1	12
Barley	11	2	60	2	14	1
Indian Corn ...	11	8	65	1	5	10
Rice	7	a trace	80	a trace	—	13
Beans and Peas	25	2	48	2	8	15
Milk	$4\frac{1}{2}$	3	5	$\frac{3}{4}$	—	$86\frac{3}{4}$

Barley is more generally used than any other grain, and, reckoned by weight, is cheaper than wheat or oats; but, unless in the form of meal, should not be the only grain given, for fowls do not fatten upon it, as, though possessing a very fair proportion of flesh-forming substances, it contains a lesser amount of fatty matters than other varieties of corn. In Surrey barley is the usual grain given, excepting during the time of incubation, when the sitting hens have oats, as being less heating to the system than the former. Barleymeal contains the same component parts as the whole grain, being ground with the husk, but only inferior barley is made into meal.

Wheat of the best description is dearer than barley, both by weight and measure, and possesses but about one-twelfth part more flesh-forming material, but it is fortunate that the small cheap wheat is the best for poultry, for Professor Johnston says, "the small or tail corn which the farmer separates before bringing his grain to market is richer in gluten (flesh-forming food) than the full-grown grain, and is therefore more nutritious." The "Henwife" finds "light wheats or tailings the best grain for daily use, and next to that barley."

Oats are dearer than barley by weight. The heaviest should be bought, as they contain very little more husk than the lightest, and are therefore cheaper in proportion. Oats and oatmeal contain much more flesh-forming material than any other kind of grain, and double the amount of fatty material than wheat, and three times as much as barley. Mowbray says oats are apt to cause scouring, and chickens become tired of them; but they are recommended by many for promoting laying, and in Kent, Sussex, and Surrey for fattening. Fowls frequently refuse the lighter samples of oats, but if soaked in water for a few hours so as to swell the kernel, they will not refuse them. The meal contains more flesh-forming material than the whole grain.

The meal of wheat and barley are much the same as the whole grain, but oatmeal is drier and separated from a large portion of the husk, which makes it too dear except for fattening fowls and feeding the youngest chickens, for which it is the very best food. Fine "middlings," also termed "sharps" and "thirds," and in London coarse country flour, are much like oatmeal, but cheaper than the best, and may be cheaply and advantageously employed instead of oatmeal, or mixed with boiled or steamed small potatoes or roots.

Many writers recommend refuse corn for fowls, and the greater number of poultry-keepers on a small scale perhaps think such light common grain the cheapest food; but this is a great mistake, as, though young fowls may be fed on offal and refuse, it is the best economy to give the older birds the finest kind of grain, both for fattening and laying, and even the young fowls should be fed upon the best if fine birds for breeding or exhibition are desired. "Instead of giving ordinary or tail corn to my fattening or breeding poultry," says Mowbray, "I have always found it most advantageous to allow the heaviest and the best; thus putting the confined fowls on a level with those at the barndoor, where they are sure to get their share of the weightiest and finest corn. This high feeding shows itself not only in the size and flesh of the fowls, but in the size, weight, and substantial goodness of their eggs, which, in these valuable particulars, will prove far superior to the eggs of fowls fed upon ordinary corn or washy potatoes; two eggs of the former going further in domestic use than three of the latter." "Sweepings" sometimes contain poisonous or hurtful substances, and are always dearer, weight for weight, than sound grain.

Some poultry-keepers recommend that the grain should be boiled, which makes it swell greatly, and consequently fills the fowl's crop with a smaller quantity, and the bird is satisfied with less than if dry grain be given; but others say that the fowls derive more nutriment from the same quantity of grain unboiled. Indeed, it seems evident that a portion of the nutriment must pass into the water, and also evaporate in steam. The fowl's gizzard being a powerful grinding mill, evidently designed by Providence for the purpose of crushing the grain into meal, it is clear that whole grain is the natural diet of fowls, and that softer kinds of food are chiefly to be used for the first or morning meal for fowls confined in houses (see p. 34), and for those being fattened artificially in coops, where it is desired to help the fowl's digestive powers, and to convert the food into flesh as quickly as possible.

Indian corn or maize, either whole or in meal, must not be given in too great a proportion, as it is very fattening from the large quantity of oil it contains; but mixed with barley or barley-meal, it is a most economical and useful food. It is useful for a change, but is not a good food by itself. It may be given once or twice a week, especially in the winter, with advantage. From its size small birds cannot eat it and rob the fowls. Whether whole or in meal, the maize should be scalded, that the swelling may be done before it is eaten. The yellow-coloured maize is not so good as that which is reddish or rather reddish-brown.

Buckwheat is about equal to barley in flesh-forming food, and is very much used on the Continent. Mr. Wright has "a strong opinion that the enormous production of eggs and fowls in France is to some extent connected with the almost universal use of buckwheat by French poultry-keepers." It is not often to be had cheap in this country, but is hardy and may be grown anywhere at little cost. Mr. Edwards says, he "obtained (without manure) forty bushels to the acre, on very poor sandy soil, that would not have produced eighteen bushels of oats. The seed is angular in form, not unlike hempseed; and is stimulating, from the quantity of spirit it contains."

Peas, beans, and tares contain an extraordinary quantity of flesh-forming material, and very little of fat-forming, but are too stimulating for general use, and would harden the muscular fibres and give too great firmness of flesh to fowls that are being fattened, but where tares are at a low price, or peas or beans plentiful, stock fowls may be advantageously fed upon any of these, and they may be given occasionally to fowls that are being fattened. It is better to give them boiled than in a raw state, especially if they are hard and dry, and the beans in particular may be too large for the fowls to swallow comfortably. Near Geneva fowls are fed chiefly upon tares. Poultry reject the wild tares of which pigeons are so fond.

Rice is not a cheap food. When boiled it absorbs a great quantity of water and forms a large substance, but, of course, only contains the original quantity of grain which is of inferior value, especially for growing chickens, as it consists almost entirely of starch, and does not contain quite half the amount of flesh-forming materials as oats. When broken or slightly damaged it may be had much cheaper, and will do as well as the finest. Boil it for half an hour in skim-milk or water, and then let it stand in the water till cold, when it will have swollen greatly, and be so firm that it can be taken out in lumps, and easily broken into pieces. In addition to its strengthening and fattening qualities rice is considered to improve the delicacy of the flesh. Fowls are especially fond of it at first, but soon grow tired of this food. If mixed with less cloying food, such as bran, they would probably continue to relish it.

Hempseed is most strengthening during moulting time, and should then be given freely, especially in cold localities.

Linseed steeped is occasionally given, chiefly to birds intended for exhibition, to increase the secretion of oil, and give lustre to their plumage.

Potatoes, from the large quantity of starch they contain, are not good unmixed, as regular food, but mixed with bran or meal are most conducive to good condition and laying. They contain a great proportion of nutriment, comparatively to their bulk and price; and may be advantageously and profitably given where the number of eggs produced is of more consequence than their flavour or goodness. A good morning meal of soft food for a few fowls may be provided daily almost for nothing by boiling the potato peelings till soft, and mashing them up with enough bran, slightly scalded, to make a tolerably stiff dry paste. The peelings will supply as many fowls as there are persons at the dinner table. A little salt should always be added, and in winter a slight sprinkling of pepper is good.

"It is indispensable," says Mr. Dickson, "to give the potatoes to fowls not only in a boiled state, but hot; not so hot, however, as to burn their mouths, as they are stupid enough to do if permitted. They dislike cold potatoes, and will not eat them willingly. It is likewise requisite to break all the potatoes a little, for they will not unfrequently leave a potato when thrown down unbroken, taking it, probably, for a stone, since the moment the skin is broken and the white of the interior is brought into view, they fall upon it greedily. When pieces of raw potatoes are accidentally in their way, fowls will sometimes eat them, though they are not fond of these, and it is doubtful whether they are not injurious."

Mangold-wurtzel, swedes, or other turnips, boiled with a very small quantity of water, until quite soft, and then thickened with the very best middlings or meal, is the very best soft food, especially for Dorkings.

Soft food should always be mixed rather dry and *friable*, and not *porridgy*, for they do not like sticky food, which clings round their beaks and annoys them, besides often causing diarrhœa. There should never be enough water in food to cause it to glisten in the light. If the soft food is mixed boiling hot at night and put in the oven, or covered with a cloth, it will be warm in the morning, in which state it should always be given in cold weather.

Fowls have their likes and dislikes as well as human beings, some preferring one kind of grain to all others, which grain is again disliked by other fowls. They also grow tired of the same food, and will thrive all the better for having as much variety of diet as possible, some little change in the food being made every few days. Fowls should not be forced or pressed to take food to which they show a dislike. It is most important to give them chiefly that which they like best, as it is a rule, with but few exceptions, that what is eaten with most relish agrees best and is most easily digested; but care must be taken not to give too much, for one sort of grain being more pleasing to their palate than another, induces them to eat gluttonously more than is necessary or healthy.

M. Réaumur made many careful experiments upon the feeding of fowls, and among them found that they were much more easily satisfied than might be supposed from the greedy voracity which they exhibit when they are fed, and that the sorts of food most easily digested by them are those of which they eat the greatest quantity.

No definite scale can be given for the quantity of food which fowls require, as it must necessarily vary with the different breeds, sizes, ages, condition, and health of the fowls; and with the seasons of the year, and the temperature of the season, much more food being necessary to keep up the proper degree of animal heat in winter than in summer; and the amount of seeds, insects, vegetables, and other food that they may pick up in a run of more or less extent. Over-feeding, whether by excess of quantity or excess of stimulating constituents, is the cause of the most general diseases, the greater proportion of these diseases, and of most of the deaths from natural causes among fowls. When fowls are neither laying well nor moulting, they should not be fed very abundantly; for in such a state over-feeding, especially with rich food, may cause them to accumulate too much fat. A fat hen ceases to lay, or nearly, while an over-fed cock becomes lazy and useless, and may die of apoplexy.

But half-fed fowls never pay whether kept for the table or to produce eggs. A fowl cannot get fat or make an egg a day upon little or poor food. A hen producing eggs will eat nearly twice as much food as at another time. In cold weather give plenty of dry bread soaked in ale.

Poultry prefer to pick their food off the ground. "No plan," says Mr. Baily, "is so extravagant or so injurious as to throw down heaps once or twice per day. They should have it scattered as far and wide as possible, that the birds may be long and healthily employed in finding it, and may not accomplish in a few minutes that which should occupy them for hours. For this reason every sort of feeder or hopper is bad. It is the nature of fowls to take a grain at a time, and to pick grass and dirt with it, which assist digestion. They should feed as pheasants, partridges, grouse, and other game do in a state of nature; if, contrary to this, they are enabled to eat corn by mouthfuls, their crops are soon overfilled, and they seek relief in excessive draughts of water. Nothing is more injurious than this, and the inactivity that attends the discomfort caused by it lays the foundation of many disorders. The advantage of scattering the food is, that all then get their share; while if it is thrown only on a small space the master birds get the greater part, while the others wait around. In most poultry-yards more than half the food is wasted; the same quantity is thrown down day after day, without reference to time of year, alteration of numbers, or variation of appetite, and that which is not eaten is trodden about, or taken by small birds. Many a poultry-yard is coated with corn and meal."

If two fowls will not run after one piece, they do not want it. If a trough is used, the best kind is the simplest, being merely a long, open one, shaped like that used for pigs, but on a smaller scale. It should be placed about a foot from one of the sides of the yard, behind some round rails driven into the ground three inches apart, so that the fowls cannot get into the troughs, so as to upset them, or tread in or otherwise dirty the food. The rails should be all of the same height, and a slanting board be fixed over the trough.

Some persons give but one meal a day, and that generally in the morning; this is false economy, for the whole of the nutriment contained in the one meal is absorbed in keeping up the animal heat, and there is no material for producing eggs. "The number of meals per day," says Mr. Wright, "best consistent with real economy will vary from two to three, according to the size of the run. If it be of moderate extent, so that they can in any degree forage for themselves, two are quite sufficient, at least in summer, and should be given early in the morning and the last thing before the birds go to roost. In any case, these will be the principal meals; but when the fowls are kept in confinement they will require, in addition, a scanty feed at mid-day. The first feeding should consist of soft food of some kind. The birds have passed a whole night since they were last fed ; and it is important, especially in cold weather, that a fresh supply should as soon as possible be got into the system, and not merely into the crop. But if grain be given, it has to be ground in the poor bird's gizzard before it can be digested, and on a cold winter's morning the delay is anything but beneficial. But, for the very same reason, at the evening meal grain forms the best food which can be supplied ; it is digested slowly, and during the long cold nights affords support and warmth to the fowls."

They should be fed at regular hours, and will then soon become accustomed to them, and not loiter about the house or kitchen door all day long, expecting food, which they will do if fed irregularly or too often, and neglect to forage about for themselves, and thus cost more for food.

Grass is of the greatest value for all kinds of poultry, and where they have no paddock, or grass-plot, fresh vegetables must be given them daily, as green food is essential to the health of all poultry, even of the very youngest chickens. Cabbage and lettuce leaves, spinach, endive, turnip-tops, turnips cut into small pieces and scattered like grain, or cut in two, radish-leaves, or any refuse, but not stale vegetables will do ; but the best thing is a large sod of fresh-cut turf. They are partial to all the mild succulent weeds, such as chickweed and *Chenopodium*, or fat-hen, and eat the leaves of most trees and shrubs, even those of evergreens ; but they reject the leaves of strawberries, celery, parsnips, carrots, potatoes, onions, and leeks. The supply of green food may be unlimited, but poultry should never be entirely fed on raw greens. Cabbage and spinach are still more relaxing when boiled than raw. They are very fond of the fruit of the mulberry and cherry trees, and will enjoy any that falls, and prevent it from being wasted.

Insect food is important to fowls, and essential for chickens and laying hens. "There is no sort of insect, perhaps," says Mr. Dickson, "which fowls will not eat. They are exceedingly fond of flies, beetles, grasshoppers, and crickets, but more particularly of every sort of grub, caterpillar, and maggot, with the remarkable exception of the caterpillar moth of the magpie (*Abraxas Grossularia*), which no bird will touch." M. Réaumur mentions the circumstance of a quantity of wheat stored in a corn-loft being much infected with the caterpillars of the small cornmoth, which spins a web and unites several grains together. A young lady devised the plan of taking some chickens to the loft to feed on the caterpillars, of which they were so fond that in a few days they devoured them all, without touching a single grain of the corn. Mr. Dickson observes, that "biscuit-dust from ships' stores, which consists of biscuit mouldered into meal, mixed with fragments still unbroken, would be an excellent food for poultry, if soaked in boiling water and given them hot. It is thus used for feeding pigs near the larger seaports, where it can sometimes be had in considerable quantity, and at a very reasonable price. It will be no detriment to this material if it be full of weevils and their grubs, of which fowls are fonder than of the biscuit itself."

There is not any food of which poultry generally are so fond as of earthworms; but all fowls are not equally fond of them, and some will not touch them. They will not eat dead worms. Too many ought not to be given, or they will become too fat and cease laying. When fowls are intended for the table worms should not be given, as they are said always more or less to deteriorate the flavour of the flesh. A good supply may easily be obtained. By stamping hard upon the ground, as anglers do, worms will rise to the surface; but a better method is to thrust a strong stake or a three-pronged potato-fork into the ground, to the depth of a foot or so, and jerk it backwards and forwards, so as to shake the soil all around. By going out with a light at night in calm, mild weather, particularly when there is dew, or after rain, a cautious observer will see large numbers of worms lying on the ground, gravel-walks, grass-plots or pastures; but they are easily frightened into their holes, though with caution and dexterity a great number, and those chiefly of the largest size, may be captured. Mr. Dickson advises that cottagers' children should be employed to imitate the example of the rooks, by following the plough or the digger, and collecting the worms which are disclosed to view; and also to collect cockchafers, "and, what would be more advantageous, they might be set to collect the grubs of this destructive insect after the plough, and thus, while providing a rich banquet for the poultry, they would be clearing the fields of a most destructive insect."

Fowls are very fond of shell snails. They are still more fattening than worms, and therefore too many must not be given when laying, but they do not injure the flavour of the flesh. Some will eat slugs, but they are not generally fond of these, and many fowls will not touch them.

One great secret of profitable poultry-keeping is, that hens cannot thrive and lay without a considerable quantity of animal food, and therefore if they cannot obtain a sufficient quantity in the form of insects, it must be supplied in meat, which, minced small, should be given daily and also to all fowls in winter, as insects are then not to be had. Mr. Baily says: "Do not give fowls meat, but always have the bones thrown out to them after dinner; they enjoy picking them, and perform the operation perfectly. Do not feed on raw meat; it makes fowls quarrelsome, and gives them a propensity to peck each other, especially in moulting time if the accustomed meat be withheld." They will peck at the wound of another fowl to procure blood, and even at their own wounds when within reach. Take care that long pieces of membrane, or thick skin, tough gristle or sinew, or pieces of bone, are not left sticking to the meat, or it may choke them, or form a lodgment in the crop. "Pieces of suet or fat," says Mr. Dickson, "are liked by fowls better than any other sort of animal food; but, if supplied in any quantity, will soon render them too fat for continuing to lay. Should there be any quantity of fat to dispose of, it ought, therefore, to be given at intervals, and mixed or accompanied with bran, which will serve to fill their crops without producing too much nutriment." It is a good plan when there are plenty of bones and scraps of meat to boil them well, and mix bran or pollard with the liquor before giving them to the fowls, as it makes the meat easier to mince, and extracts nourishment from the bones. When minced-meat is required for a large number of fowls, a mincing or sausage machine will save much time and prepare the meat better than chopping. They are as fond of fish, whether salted or fresh, as of flesh. Crumbs, fragments of pastry, and all the refuse and slops of the kitchen may be given them. Greaves, so much advertised for fowls, are very bad, rapidly throwing them out of condition, causing their feathers to fall off, spoiling the flavour of the flesh; they cause premature decrepitude, and engender many diseases, the most common being dropsy of an incurable character.

Where there is no danger from thieves, foxes, or other vermin, and the run is extensive, it is the best plan to leave the small door of the fowl-house open, and the fowls will go out at daybreak and pick up many an "early worm" and insect. The morning meal may be given when the household has risen.

A constant supply of fresh clean water is indispensable. Fountains are preferable to open vessels, in which the fowls are apt to void their dung, and the chickens to dabble and catch cold, often causing roup, cramp, &c. The simplest kind of water vessel is a saucer made of red pottery, containing several circular, concentric troughs, each about an inch wide, and of the same depth. Chickens cannot get drowned in these shallow vessels, but unless placed behind rails the water will be dirtied by the fowls. They are sold at all earthenware shops, and are used for forcing early mustard in. A capital fountain may be made with an earthenware jar or flower-pot and a flower-pot saucer. Bore a small hole in the jar or flower-pot an inch and a half from the edge of the rim, or detach a piece about three-quarters of an inch deep and one inch wide, from the rim, and if a flower-pot is used plug the hole in the bottom airtight with a piece of cork; fill the vessel with water, place the saucer bottom upwards on the top, press it closely, and quickly turn both upside down, when the water will flow into the saucer, filling up the space between it and the vessel up to the same height as the hole in the side of the jar or flower-pot, therefore the hole in the side of the rim of the vessel must not be quite so deep as the height of the side of the saucer; and above all the plug in the flower-pot must be airtight. This fountain is cheap, simple, and easily cleaned. Water may also be kept in troughs, or earthenware pans, placed in the same way. The fountains and pans should be washed and filled with fresh water once every day, and oftener in warm weather; and they should occasionally be scoured with sand to remove the green slime which collects on the surface, and produces roup, gapes, and other diseases. In winter the vessels should always be emptied at night, in order to avoid ice from forming in them, which is troublesome to remove, and snow must never be allowed to fall into them, snow-water being most injurious to poultry.

CHAPTER V.

EGGS.

During the natural process of moulting, hens cease laying because all the superabundant nutriment is required for the production of the new feathers. Fowls moult later each time; the moulting occupies a longer period, and is more severe as it becomes later, and if the weather should be cold at its termination they seldom recommence laying for some time. But young fowls moult in spring. Therefore, by having pullets and hens of different ages, and moulting at different times, a healthy laying stock may be kept up. Pullets hatched in March, and constantly fed highly, not only lay eggs abundantly in the autumn, but when killed in the following February or March, are as fat as any one could or need desire them to be, and open more like Michaelmas geese than chickens. When eggs alone are wanted, you can commence by buying in the spring as many hens as you require, and your run will accommodate, not more than a year or eighteen months old. If in good health and condition, they will be already laying, or will begin almost immediately; and, if well housed and fed, will give a constant supply of eggs until they moult in the autumn. When these hens have ceased laying, and before they lose their good condition by moulting, they should be either killed or sold, unless they are Hamburgs, Brahmas, or Cochins, and replaced by pullets hatched in March or April, which will have moulted early, and, if properly housed and fed, will begin to lay by November at the latest, and continue laying until February or March, when they may be sold or killed, being then in prime condition, and replaced as before; or, as they will not stop laying for any length of time, the best may be kept until the autumn, when, if profit is the chief consideration, they must be disposed of.* But Brahmas, Cochins, and Hamburgs will lay through the winter up to their second, or even third year. If you commence poultry-keeping in the autumn you should buy pullets hatched in the preceding spring. The best and cheapest plan of keeping up a good stock is to keep a full-feathered Cochin or two for March or April sitting; and, if necessary, procure eggs of the breed you desire. The Cochin will sit again, being only too often ready for the task; and the later-hatched chickens can be fattened profitably for the table. But if you wish to obtain eggs all the year round, and to avoid replacing of stock, or object to the trouble of rearing chickens, keep only those breeds that are non-sitters, as the Hamburgs, Polands, and Spanish; but you must purchase younger birds from time to time to keep a supply of laying hens while others are moulting.

Warmth is most essential for promoting laying. A severe frost will suddenly stop the laying of even the most prolific hens. "When," says M. Bosc, "it is wished to have eggs during the cold season, even in the dead of winter, it is necessary to make the fowls roost over an oven, in a stable, in a shed where many cattle are kept, or to erect a stove in the fowl-house on purpose. By such methods, the farmers of Ange have chickens fit for the table in the month of April, a period when they are only beginning to be hatched in the farms around Paris, although farther to the south." It is the winter management of fowls that decides the question of profit or loss, for hens will be sure to pay in the summer, even if only tolerably attended to. It is thought by many that each hen can produce only a certain number of eggs; and if such be the case, it is very advantageous to obtain a portion of them in winter when they are generally scarce and can be eaten while fresh, instead of having the whole number produced in the summer, when so many are spoiled from too long keeping in consequence of more being produced than are required for use at the time.

When the time for her laying approaches, her comb and wattles change from their previous dull hue to a bright red,

* The Practical Poultry Keeper. By Mr. L. Wright. Cassell, Petter & Galpin.

the eye brightens, the gait becomes more spirited, and sometimes she cackles for three or four days. After laying her egg on leaving the nest the hen utters a loud cackling cry, to which the cock often responds in a high-pitched kind of scream; but some hens after laying leave the nest in silence. Some hens will lay an egg in three days, some every other day, and others every day. Hens should not be forced. By unnaturally forcing a fowl with stimulating food, and more particularly with hempseed and tallow greaves, to lay in two years or so the eggs that should have been the produce of several, the hen becomes prematurely old and diseased; and it is reasonable to suppose that the eggs are not so good as they would have been if nature had been left to run its own course. The eggs ought to be taken from the nest every afternoon when no more may be expected to be laid; for if left in the nest, the heat of the hens when laying next day will tend to corrupt them.

When the shells of the eggs are somewhat soft, it is because the hens are rather inclined to grow too fat. It is then proper to mix up a little chalk in their water, and to put a little mortar rubbish in their food, the quantity of which should be diminished. We give the following remarks by an experienced poultry-keeper of the old school, as valuable from being the result of practice: "The hen sometimes experiences a difficulty in laying. In this case a few grains of salt or garlic put into the vent have been successfully tried. The keeper should indeed make use of the latter mode to find out the place where a hen has laid without his knowledge; for, as the hen will be in haste to deposit her egg, her pace towards the nest will be quickened; she may then be followed and her secret found out."

"Though one particular form," says Mr. Dickson, "is so common to eggs, that it is known by the familiar name of egg-shaped, yet all keepers of poultry must be aware that eggs are sometimes nearly round, and sometimes almost cylindrical, besides innumerable minor shades of difference. In fact, eggs differ so much in shape, that it is said experienced poultry-keepers can tell by the shape of the eggs alone the hen that laid them; for, strange to say, however different in size the eggs of any particular hen may be occasionally, they are very rarely different in form. Among the most remarkable eggs may be mentioned those of the Shanghae, or Cochin-China fowl, which are of a pale chocolate colour; and those of the Dorking fowl, which are of a pure white, and nearly as round as balls. The eggs of the Malay fowls are brown; those of the Polish fowl, which are very much pointed at one end, are of a delicate pinkish white; and those of the Bantam are of a long oval."

A very important part of the egg is the air-bag, or *folliculus æris*, which is placed at the larger end, between the shell and its lining membranes. It is, according to Dr. Paris, about the size of the eye of a small bird in new laid eggs, but enlarges to ten times that size during the process of incubation. "This air-bag," says Mr. Dickson, "is of such great importance to the development of the chick, probably by supplying it with a limited atmosphere of oxygen, that if the blunt end of the egg be pierced with the point of the smallest needle (a stratagem which malice not unfrequently suggests), the egg cannot be hatched, but perishes."

An egg exposed to the air is continually losing a portion of its moisture, the place of which is filled by the entrance of air, and the egg consequently becomes stale, and after a time putrid. M. Réaumur made many experiments in preserving eggs, and found that, by coating them with varnish, it was impossible to distinguish those which had been kept for a year from those newly laid; but varnish, though not expensive, is not always to be had in country places, and it also remained on the eggs placed under a hen and impeded the hatching, while in boiling them, the varnish, not being soluble in hot water, prevented them from being properly cooked. He tried other substances, and found that fat or grease, such as suet, lard, dripping, butter, and oil, were well adapted for the purpose, the best of these being a mixture of mutton and beef suet thoroughly melted together over a slow fire, and strained through a linen cloth into an earthen pan. It is only requisite, he says, to take a piece of the fat or butter about the size of a pea on the end of the finger, and rub it all over the shell, by passing and repassing the finger so that no part be left untouched; the transpiration of matter from the egg being as effectually stopped by the thinnest layer of fat or grease as by a thick coating, so that no part of the shell be left ungreased, or the tip of the finger may be dipped into oil and passed over the shell in the same manner. If it is desired that the eggs should look clean, they may be afterwards wiped with a towel, for sufficient grease or oil enters the pores of the shell to prevent all transpiration without its being necessary that any should be left to fill up the spaces between the pores. They can be boiled as usual without rubbing off the fat, as it will melt in the hot water, and when taken out of the water the little grease that is left upon the egg is easily wiped off with a napkin.

Eggs preserved in this manner can also be used for hatching, as the fat easily melts away by the heat of the hen; and by this means the eggs of foreign fowls might be carried to a distance, hatched, and naturalised in this and other countries. The French also find that a mixture of melted beeswax and olive oil is an excellent preservative.

Eggs may also be preserved for cooking by packing them in sawdust, in an earthen vessel, and covering the top with melted mutton suet or fat; as fruit is sometimes preserved. They are also said to keep well in salt, in a barrel arranged in layers of salt and eggs alternately. If the salt should become damp, it would penetrate through the pores of the shell and pickle them to a certain extent. M. Gagne says that eggs may be preserved in a mixture made of one bushel of quick-lime, two pounds of salt, and eight ounces of cream of tartar, with sufficient water to make it into a paste of a consistency to receive the eggs, which, it is said, may be kept in it fresh for two years; but eggs become tasteless when preserved with lime. It may be as well to mention here that eggs are comparatively wasted when used in making a rice pudding, as they render it too hard and dry, and the pudding without them, if properly made, will be just of the right consistency.

"Another way to preserve eggs," says Mr. Dickson, "is to have them cooked in boiling water the same day they are laid. On taking them out of the water they are marked with red ink, to record their date, and put away in a cool place, where they will keep, it is said, for several months. When they are wanted for use, they are again put into hot water to warm them. The curdy part which is usually seen in new-laid eggs is so abundant, and the taste is said to be so well preserved, that the nicest people may be made to believe that they are new laid. At the end of three or four months, however, the membrane lining the shell becomes much thickened, and the eggs lose their flavour. Eggs so preserved have the advantage of not suffering from being carried about."

"It ought not to be overlooked," says Mr. Dickson, "with respect to the preservation of eggs, that they not only spoil by the transpiration of their moisture and the putrid fermentation of their contents, in consequence of air penetrating through the pores of the shell; but also by being moved about, and jostled when carried to a distance by sea or land. Any sort of rough motion indeed ruptures the membranes which keep the white, the yolk, and the germ of the chick in their proper places, and upon these becoming mixed, putrefaction soon follows."

If the eggs are to be kept for setting, place a box, divided by partitions into divisions for the eggs of the different breeds, in a dry corner of your kitchen, but not too near to the fire; fill the divisions with bran previously well dried in an oven; place the eggs in it upright, with the larger ends uppermost, as soon as they are laid, and cover them with the bran. Mark each egg in pencil with the date when laid, and description of breed or cross. They should be kept in a cool place or a warm place according to the season. Airtight jars, closed with airtight stoppers, may be used if the eggs are intended to be kept for a very long time.

In selecting eggs for setting, choose the freshest, those of moderate size, well-shaped, and having the air-vessel distinctly visible, either in the centre of the top of the egg, or slightly to the side, when the egg is held between the eye and a lighted candle, in a darkened room. Reject very small eggs, which generally have no yolk, those that are ill-shaped, and those of equal thickness at both ends, which latter is the usual shape of eggs with double yolks. These should be avoided, as they are apt generally to prove unfertile, or produce monstrosities.

It has been stated that the sex of the embryo chicken can be ascertained by the position of the air-vessel; that if it be on the top the egg will produce a cockerel, and if on the side a pullet; but there is no proof of the truth of this, and, notwithstanding such assertions, it appears to be impossible to foretell the sex of the chick, from the shape of the egg or in any other way.

In selecting eggs for the purpose of producing fowls that are to be kept for laying only, being non-sitters, choose eggs only from those hens that are prolific layers, for prolific laying is often as characteristic of some fowls of a breed as it is of the particular breeds, and by careful selection this faculty, like others, may be further developed, or continued if already fully developed.

If carefully packed, eggs for setting may be carried great distances—hundreds and even thousands of miles—without injury; vibration and even moderate shaking, and very considerable changes of temperature, producing no ill effect upon the germ. The chief point is to prevent the escape of moisture by evaporation, and consequent admission of air. A hamper travels with less vibration than a box, and is therefore preferable, especially for a long journey. They should be packed in hay, by which they will be preserved from breakage much better than by being packed in short, close material like bran, chaff, oats, or sawdust; these being shaken into smaller space by the vibration of travelling, the eggs often strike and crack each other. The hamper or box should be large enough to admit of some soft, yielding packing material being placed all round the eggs. The bottom should be first covered with a good layer of hay, straw, or moss. It is a good plan to roll each egg separately in hay or moss, fastened with a little wool or worsted. They should be covered with well-rubbed straw, pressed down carefully and gently. The lid of the hamper should be sewed on tightly all round, or in three or four places at least. If a box is used, the lid should be fastened by cords or screws, but not with nails, as the hammering would probably destroy the germ of the egg.

In procuring eggs for hatching, be sure that the parent birds are of mature age, but not too old, well-shaped, vigorous, and in perfect health; that one cock is kept to every six or seven hens; and that they are well fed and attended to. Have a steady broody hen ready to take the eggs.

CHAPTER VI.

THE SITTING HEN.

ALL hens that are inclined to sit should be allowed to hatch and bring up one brood of chickens a year; for, if altogether restrained from sitting, a hen suffers much in moulting, and is restless and excited for the remainder of the season. It is unnatural, and therefore must be injurious. The period of incubation gives her rest from producing eggs. The hen that is always stimulated to produce eggs, and not allowed to vary that process by hatching and bringing up a young brood, must ultimately suffer from this constant drain upon her system, and the eggs are said to be unwholesome.

But hens frequently wish to sit when it is not convenient, or in autumn or winter, when it is not advisable, unless very late or early chickens are desired, and every attention can be given 'to them. To check this desire, the old-fashioned plan with farmers' wives, of plunging the broody hen into cold water, and keeping her there for some minutes, was not only a cruel practice, but often failed to effect its object, and must naturally always have caused ultimate disease in the poor bird. When it is absolutely necessary to check the desire of a hen to sit, the best plan is to let her sit on some nest-eggs for a week, then remove and coop her for a few days, away from the place where she made her nest, low diet, as boiled potatoes and boiled rice, and water being placed near; meanwhile taking away the eggs and destroying the nest, and, not finding it on her return, she will generally not seek for another, unless she is a Cochin, or the desire exceedingly strong.

When a hen wishes to sit, she utters a peculiar cluck, ruffles her feathers, wanders about, searches obscure corners and recesses, is very fidgety, feverishly hot, impatient, anxiously restless, and seeks for a nest. Highly-fed hens feel this desire sooner than those that are not so highly fed. A hen may be induced to sit at any season, by confining her in a dark room in a covered basket, only large enough to contain her nest, keeping her warm, and feeding her on stimulating food, such as bread steeped in ale, a little raw liver or fresh meat chopped small, and potatoes mashed warm with milk and oatmeal.

Every large poultry establishment should have a separate house for the sitting hens, and the run that should be provided for their relaxation must be divided from that of the other fowls by wire or lattice work, to prevent any intrusion. Where there is a large number of sitting hens, each nest should be numbered, and the date of setting, number and description of eggs, entered in a diary or memorandum book opposite to the number; and the number of chickens hatched, and any particulars likely to be useful on a future occasion, should afterwards be entered.

A separate house and run for each sitting hen is a great advantage, as it prevents other hens from going to the nest during her absence, or herself from returning to the wrong nest, as will often happen in a common house. The run should not be large, or the hen may be inclined to wander and stay away too long from her nest. A separate division for the sitting hen is often otherwise useful, for the purpose of keeping the cock apart from the hens, or for keeping a few additional birds for which accommodation has not been prepared, or for the use of a pen of birds about to be sent for exhibition.

"Boxes, of which every carpenter knows the form," says Mowbray, "are to be arranged round the walls, and it is proper to have a sufficient number, the hens being apt to dispute possession, and sit upon one another. The board or step at the entrance should be of sufficient height to prevent the eggs from rolling out. Provision of a few railed doors may be made for occasional use, to be hung before the entrance, in order to prevent other hens from intruding to lay their eggs upon those which sit, a habit to which some are much addicted, and by which a brood is often injured. The common deep square boxes, uncovered at top, are extremely improper, because that form obliges the hen to jump down upon her eggs, whereas for safety she should descend upon them from a very small height, or in a manner walk in upon them. The same objection lies against hampers, with the additional one of the wickerwork admitting the cold in variable weather, during winter or early spring sittings. Many breeders prefer to have all the nests upon the ground, on account of the danger of chickens falling from the nests which are placed above." The ground is preferable for other reasons. The damp arising from the ground assists very materially in incubation. When fowls sit upon wooden floors, or in boxes, the eggs become so dry and parched as to prevent the chicken from disencumbering itself of the shell, and it is liable to perish in its attempts. Hens in a state of nature make their nests upon the ground; and fowls, when left to choose a nest for themselves, generally fix upon a hedge, where the hen conceals herself under the branches of the hedge, and among the grass. In general, the sitting places are too close and confined, and very different in this respect to those that hens select for themselves.

But nests cannot always be allowed to be made on the ground, unless properly secured from vermin, particularly from rats, which will frequently convey away the whole of the eggs from under a hen. And other considerations may render it necessary to have them on a floor, in boxes on the ground, or placed above; in which cases the eggs must be kept properly moistened, for, unless the egg is kept sufficiently damp, its inner membrane becomes so hard and dry that the chicken cannot break through, and perishes. When a hen steals her nest in a hedge or clump of evergreens or bushes, she makes it on the damp ground. She goes in search of food early in the morning, before the dew is off the grass, and returns to her nest with her feathers saturated with moisture. This is the cause of the comparatively successful hatching of the eggs of wild birds. The old farmers' wives did not understand the necessity of damping eggs, but frequently complained of their not hatching, although chickens were found in them, which was, in most cases, entirely caused by want of damping. If, therefore, the weather is warm and wet, all will probably go well; but if the air should be very dry, moisture must be imparted by sprinkling the nest and eggs slightly, when the hen is off feeding, by means of a small brush dipped in tepid water. A small flat brush such as is used by painters is excellent for this purpose, as it does not distribute the water too freely. The ground round about, also, should be watered with hot water, to cause a steam. But the natural moisture of a damp soil is preferable, and never fails.

The nest may be of any shape. A long box divided by partitions into several compartments is much used, but separate boxes or baskets are preferable as being more easily cleaned and freed from vermin. Wooden nest-boxes are preferable to wicker baskets in winter, as the latter let in the cold air, but many prefer wicker baskets in summer for their airiness. A round glazed earthen pan, with shelving sides, like those used in the midland counties for milk, and partially filled with moss, forms a good nest, the moss being easier kept moist in such a pan than in a box. The nest should be made so large that the hen can just fill it, not very deep, and as nearly flat inside at the bottom as possible, so that the eggs may not lean against each other, or they may get broken, especially by the hen turning them.

The best filling for hatching nests is fine dry sand, mould, coal or wood ashes placed on a cut turf, covering it and lining the sides with a little well-broken dry grass, moss, bruised straw, lichen, or liverwort collected from trees, or dry heather, which is the best of all, but cannot always be had. Hay, though soft at first, soon becomes hard and matted, and is also said to breed vermin. Straw is good material, but must be cut into short pieces, for if long straw is used and the hen should catch her foot in it, and drag it after her when she leaves the nest, it will disturb, if not break, the eggs. The nests of the sitting hens in Her Majesty's poultry-yard at Windsor are made of heather, which offers an excellent medium between the natural damp hedge-nest of the hen and the dryness of a box filled with straw, and also enables her to free herself from those insects which are so troublesome to sitting hens. A thick layer of ashes placed under the straw in cold weather will keep in the heat of the hen. A little Scotch snuff is a good thing to keep the nests free from vermin.

Where only a few fowls are kept, and a separate place cannot be found for the sitting hen, she can be placed on a nest which should be covered over with a coop, closed in with a little boarding or some other contrivance for a day or two, to prevent her being disturbed by any other fowls that have been accustomed to lay there. They will then soon use another nest. She should be carefully lifted off her nest, by taking hold of her under the wings, regularly every morning, exercised and fed, and then shut in, so that she cannot be annoyed.

It is best to allow a hen to keep the nest she has chosen when she shows an inclination to sit; and if she continues to sit steadily, and has not a sufficient number of eggs under her, or the eggs you desire her to hatch, remove her gently at night, replace the eggs with the proper batch, and place her quietly upon the nest again. Hens are very fond of choosing their own nests in out of the way places; and where the spot is not unsafe, or too much exposed to the weather, it is best to let her keep possession, for it has been noticed that, when she selects her own nest and manages for herself, she generally brings forth a good and numerous brood. Mr. Tegetmeier observes that he has "reason to believe, indeed, that whatever care may be taken in keeping eggs, their vitality is better preserved when they are allowed to remain in the nest. Perhaps the periodical visits of the hen, while adding to her store of eggs, has a stimulating influence. The warmth communicated in the half-hour during which she occupies the nest may have a tendency to preserve the embryo in a vigorous state."

It is a good plan, before giving an untried hen choice eggs, to let her sit upon a few chalk or stale eggs for a few days, and if she continue to sit with constancy, then to give her the batch for hatching. When choice can be made out of several broody hens for a valuable batch of eggs, one should be selected with rather short legs, a broad body, large wings well furnished with feathers, and having the nails and spurs not too long or sharp. As a rule, hens which are the best layers are the worst sitters, and those with short legs are good sitters, while long-legged hens are not. Dorkings are the best sitters of all breeds, and by high feeding may be induced to sit in October, especially if they have moulted early, and with great care and attention chickens may be reared and made fit for table by Christmas. Early in the spring Dorkings only should be employed as mothers, for they remain much longer with their chickens than the Cochin-Chinas, but the latter may safely be entrusted with a brood after April. Cochins are excellent sitters, and, from the quantity of "fluff" which is peculiar to them, keep the eggs at a high and regular degree of heat. Their short legs also are advantageous for sitting. A Cochin hen can always be easily induced to sit, and eggs of theirs or of Brahma Pootras for sitting, are not wanted in the coldest weather.

Old hens are more steady sitters than pullets, more fond of their brood, and not so apt as pullets to leave them too soon. Indeed, pullets were formerly never allowed to sit before the second year of their laying, but now many eminent authorities think it best to let them sit when they show a strong desire to do so, considering that the prejudice against them upon this point is unfounded, and that young hens sit as well as older fowls. Pullets hatched early will generally begin to lay in November or December, if kept warm and well fed, and will sit in January or February.

Broody hens brought from a distance should be carried in a basket, covered over with a cloth.

The number of eggs to be set under a hen must be according to the extent of her wings and the temperature of the weather. Some say that the number may vary from nine to fourteen, but others would never give more than nine in winter and early spring, and eleven in summer, to the largest hen, and two fewer to the smaller fowls. A Cochin-China may have fifteen of her own in summer. A hen should not be allowed more eggs than she can completely cover; for eggs that are not thoroughly covered become chilled, and fewer and weaker chickens will be hatched from too large a number than from a more moderate allowance. It is not only necessary to consider how many eggs a hen can hatch, but also how many chickens she can cover when they are partly grown. In January and February, not more than seven or eight eggs should be placed under the hen, as she cannot cover more than that number of chickens when they grow large, and exposure to the cold during the long winter nights would destroy many. "The common order to set egges," says Mascall, "is in odde numbers, as seven, nyne, eleven, thirteen, &c., whiche is to make them lye round the neste, and to have the odde egge in the middest."

Eggs for sitting should be under a fortnight old, if possible, and never more than a month. Fresh eggs hatch in proper time, and, if good, produce strong, lively chicks; while stale eggs are hatched sometimes as much as two days later than new laid, and the chickens are often too weak to break the shell, while of those well out fewer will probably be reared. It is certain, as a general rule, that the older the egg the weaker will be its progeny. Every egg should be marked by a pencil or ink line drawn quite round it, so that it can be known without touching, and if another be laid afterwards it may be at once detected and removed, for hens will sometimes lay several after they have commenced sitting. Place the eggs under the hen with their larger ends uppermost.

Let the hen be well fed and supplied with water before putting her on the nest. Whole barley and soft food, chiefly barleymeal and mashed potatoes, should be given to her when she comes off the nest, and she must have as much as she will eat, for she leaves the nest but once daily, and the full heat of the body cannot be kept up without plenty of food; or she may have the same food as the general stock. A good supply of water must be always within her reach. A good-sized shallow box or pan, containing fine coal-ashes, sand, or dry earth, to cleanse herself in, should always be ready near to the nest. She should be left undisturbed, and, as far as possible, allowed to manage her own business. When a hen shows impatience of her confinement, and frequently leaves the nest, M. Parmentier advises that half only of her usual meal should be given, after which she should be replaced on the nest and fed from the hand with hemp or millet seed, which will induce her to stay constantly on her eggs. Others will sit so long and closely that they become faint for want of food. Such hens should not be fed on the nest, but gently induced with some tempting dainty to take a little exercise, for they will not leave their eggs of their own accord, and feeding on the nest has crippled many a good sitter. It is not healthy for the hen to feed while sitting on or close by the nest, for she requires a little exercise and rolling in the dust-heap, as well as that the eggs should be exposed for the air to carry off any of that stagnant vapour which M. Réaumur proved to be so destructive to the embryo chickens; and it has also been shown by physiologists that the cooling of the eggs caused by this absence of the hen is essential to allow a supply of air to penetrate through the pores of the shell, for the respiration of the chick. When there are many hens sitting at the same time, it is a good plan to take them off their nests regularly at the same time every morning to feed, and afterwards give them an opportunity to cleanse themselves in a convenient dusting-place, and, if possible, allow them exercise in a good grass run. A hen should never be caught, but driven back gently to her nest.

A good hen will not stay away more than half an hour, unless infested with vermin, from want of having a proper dust-heap. But hens have often been absent for more than an hour, and yet have hatched seven or eight chickens; and instances have been known of their being absent for five and even for nine hours, and yet hatching a few. The following remarkable instance is recorded by an excellent authority: "Eggs had been supplied and a sitting hen lent to a neighbour, and, when she had set in a granary ten days, she was shut out through the carelessness of a servant. Being a stranger in the farmyard, the hen was not recognised, but supposed to have strayed in from an adjoining walk, and thirty hours elapsed before it was discovered that the hen had left her nest. The farmer's wife despaired of her brood; but, to her surprise and pleasure, eight chickens were hatched. The tiled roof of the granary was fully exposed to the rays of the sun, and the temperature very high, probably above 80 deg. during the day, and not much lower at night." Valuable eggs, therefore, should not be abandoned on account of a rather lengthened absence; and ordinary eggs should not be discarded as worthless if the hen has already sat upon them for a fortnight or so; but if she has been sitting for only a few days, it is safer to throw them away, and have a fresh batch.

During the hen's absence, always look at the eggs, remove any that may have been broken, and very gently wash any sticky or dirty eggs with a flannel dipped in milk-warm water. See that they are dry before putting them back. If the nest is also dirty, replace it with fresh material of the same kind. Gently drive the hen back to her nest as quickly as possible, to prevent any damage from the eggs becoming chilled. If a hen should break an egg with her feet or otherwise, it should be removed as soon as it is seen, or she may eat it, and, liking the taste, break and eat the others. Some hens have a bad habit of breaking and eating the eggs on which they are sitting, to cure which some recommend to boil an egg hard, bore a few holes in it, so that the inside can be seen, and give it while hot to the culprit, who will peck at the holes and burn herself; but hens with such propensities should be fattened for the table, for they are generally useless either for sitting or laying.

Some persons examine the eggs after the hen has sat upon them for six or seven days, and remove all that are sterile, by which plan more warmth and space are gained for those that are fertile, and the warmth is not wasted upon barren eggs. They may be easily proved by holding them near to the flame of a candle, the eye being kept shaded by one hand, when the fertile eggs will appear dark and the sterile transparent. Another plan is to place the eggs on a drum, or between the hands, in the sunshine, and observe the shadow. If this wavers, by the motion of the chick, the eggs are good; but if the shadow shows no motion, they are unfertile. If two hens have been sitting during the same time, and many unfertile eggs are found in the two nests, all the fertile eggs should be placed under one hen, and a fresh batch given to the other. The eggs should not be moved after this time, except by the hen, more especially when incubation has proceeded for some time, lest the position of the chick be interfered with, for if taken up a little time before its exit, and incautiously replaced with the large end lowermost, the chicken, from its position, will not be able to chip the shell, and must therefore perish. The forepart of the chicken is towards the biggest end of the egg, and it is so placed in the shell that the beak is always uppermost. When the egg of a choice breed has been cracked towards the end of the period of incubation, the crack may be covered with a slip of gummed paper, or the unprinted border that is round a sheet of postage stamps, and the damaged egg will probably yet produce a fine chick.

It is a good plan to set two hens on the same day, for the two broods may be united under one if desirable; and on the hatching day, to prevent the newly-born chickens being crushed by the unhatched eggs, all that are hatched can be given to one hen, and the other take charge of the eggs, which are then more likely to be hatched, as, while the chickens are under the hen, she will sit higher from the eggs, and afford them less warmth when they require it most.

The hen of all kinds of gallinaceous fowls, from the Bantam to the Cochin-China, sits for twenty-one days, at which time, on an average, the chickens break the shell; but if the eggs are new laid it will often lessen the time by five or six hours, while stale eggs will always be behind time. For the purpose of breaking the shell, the yet soft beak of the chicken is furnished, just above the point of the upper mandible, with a small, hard, horny scale, which, from the position of the head, as Mr. Yarrell observes, is brought in contact with the inner surface of the shell. This scale may be always seen on the beaks of newly-hatched chickens, but in the course of a short time peels off. It should not be removed. The peculiar sound, incorrectly called "tapping," so perceptible within the egg about the nineteenth day of incubation, which was universally believed to be produced by the bill of the chick striking against the shell in order to break it and effect its release, has been incontestably proved, by the late Dr. F. R. Horner, of Hull, in a paper read by him before the British Association for the Advancement of Science, to be a totally distinct sound, being nothing more than the natural respiratory sound in the lungs of the young chick, which first begins to breathe at that period. Of course there is also an occasional sound made by the tapping of the beak in endeavouring to break the shell.

The time occupied in breaking the shell varies, according to the strength of the chick, from one to three hours usually, but extends sometimes to twenty-four, and even more. "I have seen," says Réaumur, "chicks continue at work for two days together; some work incessantly, while others take rest at intervals, according to their physical strength. Some, I have observed, begin to break the shell a great deal too soon; for, be it observed, they ought, before they make their exit, to have within them provision enough to serve for twenty-four hours without taking food, and for this purpose the unconsumed portion of the yolk enters through the navel. The chick, indeed, which comes out of the shell without taking up all the yolk is certain to droop and die in a few days after it is hatched. The assistance which I have occasionally tried to give to several of them, by way of completing their deliverance, has afforded me an opportunity of observing those which had begun to break their shells before this was accomplished;

and I have opened many eggs much fractured, in each of which the chick had as yet much of the yolk not absorbed. Some chicks have greater obstacles to overcome than others, since all shells are not of an equal thickness nor of an equal consistence; and the same inequality takes place in the lining membrane, and offers still greater difficulty to the emergent chick. The shells of the eggs of birds of various species are of a thickness proportionate to the strength of the chick that is obliged to break through them. The canary-bird would never be able to break the shell it is enclosed in if that were as thick as the egg of a barndoor fowl. The chick of a barndoor fowl, again, would in vain try to break its shell if it were as thick and hard as that of an ostrich; indeed, though an ostrich ready to be hatched is perhaps thrice as large as the common chick, it is not easy to conceive how the force of its bill can be strong enough to break a shell thicker than a china cup, and the smoothness and gloss of which indicate that it is nearly as hard—sufficiently so to form, as may be often seen, a firm drinking-cup. It is the practice in some countries to dip the eggs into warm water at the time they are expected to chip, on the supposition that the shell is thereby rendered more fragile, and the labour of the chick lightened. But, though the water should soften it, upon drying in the air it would become as hard as at first. When the chick is entirely or almost out of the shell, it draws its head from under its wing, where it had hitherto been placed, stretches out its neck, directing it forwards, but for several minutes is unable to raise it. On seeing for the first time a chick in this condition, we are led to infer that its strength is exhausted, and that it is ready to expire; but in most cases it recruits rapidly, its organs acquire strength, and in a very short time it appears quite another creature. After having dragged itself on its legs a little while, it becomes capable of standing on them, and of lifting up its neck and bending it in various directions, and at length of holding up its head. At this period the feathers are merely fine down, but, as they are wet with the fluid of the egg, the chick appears almost naked.

From the multitude of their branchlets these down feathers resemble minute shrubs; when, however, these branchlets are wet and sticking to each other, they take up but very little room; as they dry they become disentangled and separated. The branchlets, plumules, or beards of each feather are at first enclosed in a membranous tube, by which they are pressed and kept close together; but as soon as this dries it splits asunder, an effect assisted also by the elasticity of the plumules themselves, which causes them to recede and spread themselves out. This being accomplished, each down feather extends over a considerable space, and when they all become dry and straight, the chick appears completely clothed in a warm vestment of soft down."

If they are not out in a few hours after the shell has been broken, and the hole is not enlarged, they are probably glued to the shell. Look through the egg then, and, if all the yolk has passed into the body of the chicken, you may assist it by enlarging the fracture with a pair of fine scissors, cutting up towards the large end of the egg, never downwards. "If," says Miss Watts, "the time has arrived when the chicken may with safety be liberated, there will be no appearance of blood in the minute blood-vessels spread over the interior of the shell; they have done their work, and are no longer needed by the now fully developed and breathing chick.. If there should be the slightest appearance of blood, resist at once, for its escape would generally be fatal. Do not attempt to let the chicken out at once, but help it a little every two or three hours. The object is not to hurry the chicken out of its shell, but to prevent its being suffocated by being close shut up within it. If the chick is tolerably strong, and the assistance needful, it will aid its deliverance with its own exertions." When the chicken at last makes its way out, do not interfere with it in any way, or attempt to feed it. Animal heat alone can restore it. Weakness has caused the delay, and this has probably arisen from insufficient warmth, perhaps from the hen having had too many eggs to cover thoroughly, or they may have been stale when set. Should you have to assist it out of the shell, take it out gently with your fingers, taking great care not to tear any of its tender skin, when freeing the feathers from the shell.

Mr. Wright says: "We never ourselves now attempt to assist a chick from the shell. If the eggs were fresh, and proper care has been taken to preserve moisture during incubation, no assistance is ever needed. To fuss about the nest frets the hen exceedingly; and we have always found that, even where the poor little creature survived at the time, it never lived to maturity. Should the reader attempt such assistance, in cases where an egg has been long chipped, and no further progress made, let the shell be cracked gently all round, without tearing the inside membrane; if that be perforated, the viscid fluid inside dries and glues the chick to the shell. Should this happen, or should both shell and membrane be perforated at first, introduce the point of a pair of scissors and cut up the egg towards the large end, where there will be an empty space, remembering that, if blood flows, all hope is at end. Then put the chick back under the hen; she will probably squeeze it to death, it is true—it is so very weak; but it will never live if put by the fire, at least we always found it so. Indeed, as we have said, we consider it quite useless to make the attempt at all."

The fact is, it is scarcely worth while to attempt to assist in the case of ordinary eggs, but if the breed is valuable the labour may be well bestowed.

Some hens are reluctant to give up sitting, and will hatch a second brood with evident pleasure; but it is cruel to overtask their strength and patience, and they are sure to suffer, more or less, from the unnatural exertion.

Some breeders use a contrivance called an "artificial mother" for broods hatched under the hen, and it may be employed very advantageously when any accident has happened to her. It is made in various forms, such as a wooden frame, or shallow box, open at both ends, and sloping like a writing-desk, with a perforated lid lined with sheep or lamb's skin, goose-down, or some similar warm fleecy material hanging down, under and between which the chickens nestle, heat being applied to the lid either by hot water or hot air, so as to imitate the warmth of the hen's breast. When chickens are hatched by artificial means, such as by the Hydro-Incubator, or the Eccaleobion, or in an oven according to the method practised by the Egyptians, these protectors are essential; for without a good substitute for the hen's natural warmth the chickens would perish. Artificial incubators are now extensively used, and where gas is laid on they are easily managed, but the chief difficulty is in rearing the chickens. For information on the subject see the works of Tegetmeier, Dickson, and Wright, on Poultry.

CHAPTER VII.

REARING AND FATTENING FOWLS.

THE first want which the chick will feel will be that of warmth, and there is no warmth so suited for them as that of the hen's body. Some persons remove the chicks from under the mother as soon as they are hatched, one by one, placing them in a basket covered up with flannel, and keep them there in a warm place, until the last chick is out, when they are put back under the hen. But this is very seldom necessary unless the weather is very cold and the hen restless, and is generally more likely to annoy than benefit her. Nor should the hen be induced to leave the nest, but be left undisturbed until she leaves of her own accord, when the last hatched chickens will be in a better condition to follow her than if she had been tempted to leave earlier. In a few hours they are able to run about and follow their parent; they do not require to be fed in the nest like most birds, but pick up the food which their mother shows them; and repose at night huddled up beneath her wings. The chicken during its development in the egg is nourished by the yolk, and the remaining portion of the yolk passes into its body previous to its leaving the shell, being designed for its first nourishment; and the chicken, therefore, does not require any food whatever during the first day. The old-fashioned plan, so popular with "practical" farmers' wives, of cramming a peppercorn down the throat of the newly-hatched chick is absurd and injurious.

The first food must be very light and delicate, such as crumbs of bread soaked in milk, the yolk of an egg boiled hard, and curds; but very little of anything at first except water, for thirst will come before hunger. The thirsty hen will herself soon teach the little ones how to drink. If your chicks be very weakly, you may cram them with crumbs of good white bread, steeped in milk or wine, but at the same time recollect that their little craws are not capable of holding more than the bulk of a pea; so rather under than over feed them.

As soon as the hen leaves the nest, she should have as much grain as she can eat, and a good supply of pure, clean water. In winter, or settled wet weather, she should, if possible, be kept on her nest for a day, and, when removed, be cooped in a warm, dry shed or outhouse; but in summer, if the weather be fine, and the chickens well upon their legs, they may be at once cooped out in the sun, on dry gravel, or if possible on a nice grass-plot, with food and water within her reach. The hen is cooped to prevent her from wearying the brood by leading them about until they are over-tired, besides being exposed to danger from cats, hawks, and vermin, tumbling into ditches, or getting wet in the high grass. They can pass in and out between the bars of the coop, and will come when she calls, or they wish to shelter under her wings. It is a good plan to place the coop for the first day out upon some dry sand, so that the hen can cleanse herself comfortably. The common basket coop should only be used in fine weather, and some straw, kept down by a stone, matting, or other covering, should be placed on the t p, to shelter them from the midday sun; otherwise a wooden coop should be used, open in front only, about two and a half or three feet square; well-made of stout, sound boards, with a gabled roof covered with felt; and at night a thick canvas or matting should be hung over the front, sufficient space being left for proper ventilation, but ot to admit cold draught, or to allow the chicks to get out. Mr. Wright describes an excellent coop which is "very common in some parts of France, and consists of two compartments, separated by a partition of bars, one ompartment being closed in front, the other fronted with bars like the partition. Each set of bars should have a sli ling one to serve as a door, and the whole coop should be t'ght and sound. It is best to have no bottom, but to put it on loose dry earth or ashes, an inch or two deep. Each half of the coop is about two feet six inches square, and may or may not be lighted from the top by a small pane of glass. The advantage of such a coop is that, except in very severe weather, no further shelter is required, even at night [if placed under a shed]. During the day the hen is kept in the outer compartment, the chickens having liberty, and the food and water being placed outside; whilst at night she is put in the inner portion of the coop, and a piece of canvas or sacking hung over the bars of the outer half. If the top be glazed, a little food and the water-vessel may be placed in the outer compartment at night, and the chicks will be able to run out and feed early in the morning, being prevented by the canvas from going out into the cold air. It will be only needful to remove the coop every two days for a few minutes, to take away the tainted earth and replace it with fresh. There should, if possible, be a grass-plot in front of the shed, the floor of which should be covered with dry, loose dust or earth." The hen should be kept under a coop until the brood has grown strong. Some breeders object to cooping, on account of its preventing the hen from scratching for worms and insects for her brood, and which are far superior to the substitutes with which they must be supplied, unless, indeed, a good supply of worms, ants' eggs, insects, or gentles can be had. The hen too has not sufficient exercise after her long sitting. Cooping thus has its advantages and disadvantages, and its adoption or not should depend upon circumstances. If it is preferred not to coop the hen, and she should be inclined to roam too far, a small run may be made with network, or with the moveable wirework described on page 21.

Winter-hatched chickens must be reared and fed in a warm place, which must be kept at an equal temperature. They return a large profit for the great care they require in hatching and rearing.

Chickens should be fed very often; every two hours is not too frequently. The number of these meals must be reduced by degrees to four or five, which may be continued until they are full grown. Grain should not be given to newly-hatched chickens. The very best food for them, after their first meal of bread-crumbs and egg, is made of two parts of coarse oatmeal and one part of barley-meal, mixed into a thick crumbly paste with milk or water. If milk is used, it must be fresh mixed for each meal, or it will become sour. Cold oatmeal porridge is an excellent food, and much liked by them. After the first week they may have cheaper food, such as bran, oatmeal, and Indian meal mixed, or potatoes mashed with bran. In a few days they may also have some whole grain, which their little gizzards will then be fully able to grind. Grits, crushed wheat, or bruised oats, should form the last meal at night. Bread sopped in water is the worst food they can have, and even with milk is still inferior to meal. For the first three or four days they may also have daily the yolk of an egg boiled hard and chopped up small, which will be sufficient for a dozen chicks; and afterwards, a piece of cooked meat, rather underdone, the size of a good walnut, minced fine, should be daily given to the brood until they are three weeks old. In winter and very early spring this stimulating diet may be given regularly, and once a day they should also have some stale bread soaked in ale; and whenever chickens suffer from bad feathering, caused either by the coldness of the season or delicacy of constitution, they must be fed highly, and have a daily supply of bread soaked in ale. Ants' eggs, which are well known as the very best animal diet for young pheasants, are also excellent for young chickens; and when a nest can be obtained it should be thrown with its surrounding mould into the run for them to peck at. Where there is no grass-plot they should have some grass cut into small pieces, or other vegetable food minced small, until they are able to peck pieces from the large leaves. Onion tops and leeks chopped small, cress, lettuce, and cabbage, are much relished by all young poultry. The French breeders give a few dried nettle seeds occasionally. Young growing fowls can scarcely have too much food, so long as they eat it with a good appetite, and do not tread any about, or otherwise leave it to waste.

Young poultry cannot thrive if overcrowded. They should not be allowed to roost on the branches of trees or shrubs, or otherwise out of doors, even in the warmest weather, or they will acquire the habit of sleeping out, which cannot be easily overcome; not that they would suffer much from even severe weather, when once accustomed to roosting out of doors, but from want of warmth the supply of eggs would decrease, and it would, in many places, be unsafe and, in most, inconvenient.

The sooner chickens can be fattened, of course the greater must be the profit. They should be put up for fattening as soon as they have quitted the hen, for they are then generally in good condition, but begin to lose flesh as their bones develop and become stronger, particularly those fowls which stand high on the leg.

Fowls are in perfection for eating just before they are fully developed. By keeping young fowls, especially the cockerels, too long before fattening them for market or home consumption, they eat up all the profit that would be made by disposing of them when the pullets have ceased laying just before their first adult moult, and the cockerels before their appetites have become large. Fowls intended to be fattened should be well and abundantly fed from their birth; for if they are badly fed during their growth they become stunted, the bones do not attain their full size, and no amount of feeding will afterwards supply these defects and transform them into fine, large birds. Poultry that have been constantly fed well from their birth will not only be always ready for the table, with very little extra attention and feeding, but their flesh will be superior in juiciness and rich flavour to those which are fattened up from a poor state. In choosing full-grown fowls for fattening, the short-legged and early-hatched should be preferred.

In fattening poultry, "the well-known common methods," Mowbray observes, "are, first, to give fowls the run of the farmyard, where they thrive upon the offals of the stables and other refuse, with perhaps some small regular feeds; but at threshing time they become fat, and are thence styled barndoor fowls, probably the most delicate and high-flavoured of all others, both from their full allowance of the finest corn and from the constant health in which they are kept, by living in the natural state, and having the full enjoyment of air and exercise; or secondly, they are confined during a certain number of weeks in coops; those fowls which are soonest ready being drawn as wanted." "The former method," says Mr. Dickson, "is immeasurably the best as regards the flavour and even wholesomeness of the fowls as food, and though the latter mode may, in some cases, make the fowls fatter, it is only when they have been always accustomed to confinement; for when barndoor fowls are cooped up for a week or two under the notion of improving them for the table, and increasing their fat, it rarely succeeds, since the fowls generally pine for their liberty, and, slighting their food, lose instead of gaining additional flesh."

To fatten fowls that have not the advantage of a barndoor, Mowbray recommends fattening-houses large enough to contain twenty or thirty fowls, warm and airy, with well-raised earth floors, slightly littered down with straw, which should be often changed, and the whole place kept perfectly clean. "Sandy gravel," he says, "should be placed in several different layers, and often changed. A sufficient number of troughs for both water and food should be placed around, that the stock may feed with as little interruption as possible from each other, and perches in the same proportion should be furnished for those birds which are inclined to perch, which few of them will desire after they have begun to fatten, but it helps to keep them easy and contented until that period. In this manner fowls may be fattened to the highest pitch, and yet preserved in a healthy state, their flesh being nearly equal in quality to the barndoor fowl. To suffer fattening fowls to perch is contrary to the general practice, since it is supposed to bend and deform the backbone; but as soon as they become heavy and indolent from feeding, they will rather incline to roost in the straw, and the liberty of perching has a tendency to accelerate the period when they wish for rest."

The practice of fattening fowls in coops, if carried to a moderate extent, is not objectionable, and may be necessary in many cases. The coop may be three feet high, two feet wide, and four feet long, which will accommodate six or eight birds, according to their size; or it may be constructed in compartments, each being about nine inches by eighteen, and about eighteen inches high. The floor should not consist of board, but be formed of bars two inches wide, and placed two inches apart. The bars should be laid from side to side, and not from the back to the front of the coop. They should be two inches wide at the upper part, with slanting or rounded sides, so as to prevent the dung from sticking to them instead of falling straight between. The front should be made of rails three inches apart. The house in which the coops are placed should be properly ventilated, but free from cold draughts, and kept of an even temperature, which should be moderately warm. The fronts of the coops should be covered with matting or other kind of protection in cold weather. The coop should be placed about two inches from the ground, and a shallow tray filled with fresh dry earth should be placed underneath to catch the droppings, and renewed every day.

When fowls are put up to fat they should not have any food given to them for some hours, and they will take it then more eagerly than if pressed upon them when first put into the coop. But little grain should be given to fowls during the time they are fattening in coops; indeed the chief secret of success consists in supplying them with the most fattening food without stint, in such a form that their digestive mills shall find no difficulty in grinding it. Buckwheat-meal is the best food for fattening; and to its use the French, in a great measure, owe the splendid condition of the fowls they send to market. If it cannot be had, the best substitute is an equal mixture of maize-meal and barley-meal. The meal may be mixed with skim milk if available. Oatmeal and barley-meal alternately, mixed with milk, and occasionally with a little dripping, is good fattening food. Milk is most excellent for all young poultry. A little chopped green food should be given daily, to keep their bowels in a proper state.

The feeding-troughs, which must be kept clean by frequent scouring, should be placed before the fowls at regular times, and when they have eaten sufficient it is best to remove them, and place a little gravel within reach to assist digestion. Each fowl should have as much food as it will eat at one time, but none should be left to become sour. A little barley may, however, be scattered within their reach. A good supply of clean water must be always within their reach. If a bird appears to be troubled with vermin, some powdered sulphur, well rubbed into the roots of the feathers, will give immediate relief. The coops should be thoroughly lime-washed after the fowls are removed, and well dried before fresh birds are put up in them.

It is a common practice to fatten poultry in coops by a process called "cramming," by which they are loaded with greasy fat in a very short time. But it is evident that such overtaxing of the fowls' digestive powers, want of exercise and fresh air, confinement in a small space, and partial deprivation of light, without which nothing living, either animal or vegetable, can flourish, cannot produce healthy or wholesome flesh. "Indeed," as Mowbray observes, "it seems contrary to reason, that fowls fed upon such greasy, impure mixtures can possibly produce flesh or fat so firm, delicate, high-flavoured, or nourishing, as those fattened upon more simple and substantial food; as for example, meal and milk, and perhaps either treacle or sugar. With respect to grease of any kind, its chief effect must be to render the flesh loose and of a coarse flavour. Neither can any advantage be gained, except perhaps a commercial one, by very quick feeding; for real excellence cannot be obtained but by waiting nature's time, and using the best food. Besides all this, I have been very unsuccessful in my few attempts to fatten fowls by cramming; they seem to loathe the crams, to pine, and to lose the flesh they were put up with, instead of acquiring flesh; and when crammed fowls do succeed, they must necessarily, in the height of their fat, be in a state of disease." Mr. Muirhead, poulterer to Her Majesty in Scotland, says: "With regard to *cramming*, I may say that it is *wholly* unnecessary, provided the fowls have abundance of the best food at regular intervals, fresh air, and a free run; in confinement fowls may gain fat, but they lose flesh. None but those who have had experience can form any idea how both qualities can be obtained in a natural way. I have seen fowls reared at Inchmartine (which had never been shut up, or had food forced upon them), equal, if not superior, to the finest Surrey fowl, or those fattened by myself for the Royal table."

If "cramming" is practised it should be done in the following manner: The feeder, usually a female, should take the fowl carefully out of the fatting-coop by placing both her hands gently under its breast, then sit down with the bird upon her lap, its rump under her left arm, open its mouth with the finger and thumb of the left hand, take the pellet with the right, dip it well into water, milk, or pot liquor, shake the superfluous moisture from it, put it into the mouth, "cram" it gently into the gullet with her forefinger, then close the beak and gently assist it down into the crop with the forefinger and thumb, without breaking the pellet, and taking great care not to pinch the throat. When the fowl has been "crammed" it should be carefully carried back to its coop, both hands being placed under its breast as before. Chickens should be "crammed" regularly every twelve hours. The "cramming" should commence with a few pellets, and the number be gradually increased at each meal until it amounts to about fifteen. But always before you begin to feed gently feel the fowl's crop to ascertain that the preceding meal has been digested, and if you find it to contain food, let the bird wait until it is all digested, and give it fewer pellets at the next meal. If the "crams" should become hardened in the crop, some lukewarm water must be given to the bird, or poured down its throat if disinclined to drink, and the crop be gently pressed with the fingers until the hardened mass has become loosened so that the gizzard can grind it.

The food chiefly used in France for "cramming" fowls is buckwheat-meal bolted very fine and mixed with milk. It should be prepared in the following manner: Pour the milk, which should be lukewarm in winter, into a hole made in the heap of meal, mixing it up with a wooden spoon a little at a time as long as the meal will take up the milk, and make it into the consistency of dough, keep kneading it until it will not stick to the hands, then divide it into pieces twice as large as an egg, which form into rolls generally about as thick as a small finger, but more or less thick according to the size of the fowls to be fed, and divide the rolls into pellets about two and a half inches in length by a slanting cut, which leaves pointed ends, that are easier to "cram" the fowls with than if they were square. The pellets should be rolled up as dry as possible.

The operation of caponising as performed in England is barbarous, extremely painful, and dangerous. In France it is performed in a much more scientific and skilful manner. But the small advantage gained by this unnatural operation is more than counterbalanced by the unnecessary pain inflicted on the bird, and the great risk of losing it. Capons never moult, and lose their previously strong, shrill voice. In warm, dry countries they grow to a large size, and soon fatten, but do not succeed well in our moist, cold climate. They are not common in this country, and most of the fowls sold in the London markets as capons are merely young cockerels well crammed. If capons are kept they should have a separate house, for the other fowls will not allow those even of their own family to occupy the same roosting-perch with them. The hens not only show them indifference, but decided aversion. Hen chickens, deprived of their reproductive organs in order to fatten them sooner, are common in France, where they are styled poulardes.

Fattening ought to be completed in from **ten to twenty**

days. When fowls are once fattened up they should be killed, for they cannot be kept fat, but begin to lose flesh and become feverish, which renders their flesh red and unsaleable, and frequently causes their death.

Great cruelty is often ignorantly inflicted by poulterers, higglers, and others, in "twisting the necks" of poultry. An easy mode of killing a fowl is to give the bird a very sharp blow with a small but heavy blunt stick, such as a child's bat or wooden sword, at the back of the neck, about the second or third joint from the head, which will, if properly done, sever the spine and cause death very speedily. But the knife is the most merciful means; the bird being first hung up by the legs, the mouth must be opened wide, and a long, narrow, sharp-pointed knife, like a long penknife, which instrument is made for the purpose, should be thrust firmly through the back part of the roof of the mouth up into the brain, which will cause almost instant death. Another mode of killing is to pluck a few feathers from the side of the head, just below the ear, and make a deep incision there. Some say that fowls should not be bled to death like turkeys and geese, as, from the loss of blood, the flesh becomes dry and insipid. But when great whiteness of flesh is desired, the fowl should be hung up by its legs immediately after being killed, and if it has been killed without the flow of blood, an incision should be made in the neck so that it may bleed freely.

Fowls that have been kept without food and water for twelve hours before being killed will keep much better than if they had been recently fed, as the food is apt to ferment in the crop and bowels, which often causes the fowl to turn green in a few hours in warm weather. If empty they should not be drawn, and they will keep much better. Fowls are easiest plucked at once, while warm; they should afterwards be scalded by dipping them for a moment in boiling water, which will give a plump appearance to any good fowl. Fowls should not be packed for market before they are quite cold. Old fowls should not be roasted, but boiled, and they will then prove tolerably good eating.

The feathers are valuable and should be preserved, which is very easily managed. "Strip the plumage," says Mr. Wright, "from the quills of the larger feathers, and mix with the small ones, putting the whole loosely in paper bags, which should be hung up in the kitchen, or some other warm place, for a few days to dry. Then let the bags be baked three or four times for half an hour each time, in a cool oven, drying for two days between each baking, and the process will be completed. Less trouble than this will do, and is often made to suffice; but the feathers are inferior in crispness to those so treated, and may occasionally become offensive."

CHAPTER VIII.

STOCK, BREEDING, AND CROSSING.

KEEP only good, healthy, vigorous, well-bred fowls, whether you keep them to produce eggs or chickens, or both. The ill-bred mongrel fowls which are so commonly kept, are the most voracious, and consume larger quantities of food, without turning it to any account; while well-bred fowls eat less, and quickly convert that into fat, flesh, and eggs. "Large, well-bred fowls," says Mr. Edwards, " do not consume more food than ravenous, mongrel breeds. It is the same with fowls as with other stock. I have at this moment two store pigs, one highly bred, the other a rough, ill-bred animal. They have, since they left their mothers, been fed together and upon the same food. The former, I am confident (from observation), ate considerably less than the latter, which was particularly ravenous. The former pig, however, is in excellent condition, kind, and in a measure fat; whereas the latter looks hard, starved, and thin, and I am sure she will require one-third more food to make bacon of."

For the amateur who is content with eggs and chickens, and does not long for prize cups, excellent birds possessing nearly all the best characteristics of their breeds, but rendered imperfect by a few blemishes, may be purchased at a small cost, and will be as good layers or chicken-producers, and answer his purpose as well as the most expensive that can be bought.

The choice of breed must depend upon the object for which the fowls are kept, whether chiefly for eggs or to produce chickens, or for both; the climate, soil, and situation; the space that can be allotted to them; and the amount of attention that can be devoted to their care. If fowls are to be bred for exhibition, you must be guided by your own taste, pocket, and resources, as well as by the suitability of the situation for the particular breed desired. The advantages, disadvantages, and peculiarities of the various breeds will be described under their respective heads.

In commencing poultry-keeping buy only young and healthy birds. No one sign is infallible to the inexperienced. In general, however, the legs of a young hen look delicate and smooth, her comb and wattles are soft and fresh, and her general outline, even in good condition (unless when fattened for the table), rather light and graceful; whilst an old one will have rather hard, horny-looking shanks; her comb and wattles look somewhat harder, drier, and more "scurfy," and her figure is well filled out. But any of these signs may be deceptive, and the beginner should use his own powers of observation, and try and catch the "old look," which he will soon learn to know.

All authorities agree that a cock is in his prime at two years of age, though some birds show every sign of full vigour when only four months old. It is agreed by nearly all the greatest authorities that the ages of the cocks and hens should be different; however, good birds may be bred from parents of the same age, but they should not be less than a year old. The strongest chickens are obtained from two-year old hens by a cockerel of about a year old; but such broods contain a disproportion of cocks, and, therefore, most poultry-keepers prefer to breed from well-grown pullets of not less than nine months with a cock of two years of age. The cock should not be related to the hens. It is, therefore, not advisable to purchase him from the same breeder of whom you procure the hens. Do not let him be the parent of chickens from pullets that are his own offspring. Breeding in-and-in causes degeneracy in fowls as in all other animals. Some birds retain all their fire and energy until five or even six years of age, but they are beyond their prime after the third, or at the latest their fourth year; and should be replaced by younger birds of the same breed, but from a different stock.

Poultry-breeders differ with respect to the proper

number of hens that should be allowed to one cock. Columella, who wrote upon poultry about two thousand years ago, advised twelve hens to one cock, but stated that "our ancestors did use to give but five hens." Stephanus gave the same number as Columella. Bradley, and the authors of the "Complete Farmer," and the article upon the subject in "Rees's Cyclopædia," give seven or eight; and those who breed game-cocks are particular in limiting the number of hens to four or five for one cock, in order to obtain strong chickens. If fine, strong chickens be desired for fattening or breeding, there should not be more than five or six hens to one cock; but if the supply of eggs is the chief consideration, ten or twelve may be allowed; indeed, if eggs are the sole object, he can be dispensed with altogether, and his food saved, as hens lay, if there be any difference, rather better without one.

The russet red is the most hardy colour, white the most delicate, and black the most prolific. General directions for the choice of fowls, as to size, shape, and colour, cannot be applicable to all breeds, which must necessarily vary upon these points. But in all breeds the cock should, as M. Parmentier says, "carry his head high, have a quick, animated look, a strong, shrill voice (except in the Cochins, which have a fuller tone), a fine red comb, shining as if varnished, large wattles of the same colour, strong wings, muscular thighs, thick legs furnished with strong spurs, the claws rather bent and sharply pointed. He ought also to be free in his motions, to crow frequently, and to scratch the ground often in search of worms, not so much for himself, as to treat his hens. He ought, withal, to be brisk, spirited, ardent, and ready in caressing the hens, quick in defending them, attentive in soliciting them to eat, in keeping them together, and in assembling them at night."

To prevent cocks from fighting, old Mascall, following Columella, says: "Now, to slacke that heate of jealousie, ye shall slitte two pieces of thicke leather, and put them on his legges, and those will hang over his feete, which will correct the vehement heate of jealousie within him"; and M. Parmentier observes that "such a bit of leather will cause the most turbulent cock to become as quiet as a man who is fettered at the feet, hands, and neck."

The hen should be of good constitution and temper, and, if required to sit, large in the body and wide in the wings, so as to cover many eggs and shelter many chickens, but short in the legs, or she could not sit well. M. Parmentier advises the rejection of savage, quarrelsome, or peevish hens, as such are seldom favourites with the cocks, scarcely ever lay, and do not hatch well; also all above four or five years of age, those that are too fat to lay, and those whose combs and claws are rough, which are signs that they have ceased to lay. Hens should not be kept over their third year unless very good or choice. Hens are not uncommon with the plumage and spurs of the cock, and which imitate, though badly, his full-toned crow. In such fowls the power of producing eggs is invariably lost from internal disease, as has been fully demonstrated by Mr. Yarrell in the "Philosophical Transactions" for 1827, and in the "Proceedings of the Zoological Society" for 1831. Such birds should be fattened and killed as soon as observed.

By careful study of the characteristics of the various breeds, breeding from select specimens, and judicious crossing, great size may be attained, maturity early developed, facility in putting on flesh encouraged, hardiness of constitution and strength gained, and the inclination to sit or the faculty of laying increased.

Sir John Sebright, speaking of breeding cattle, says: "Animals may be said to be improved when any desired quality has been increased by art beyond what that quality was in the same breed in a state of nature. The swiftness of the racehorse, the propensity to fatten in cattle, and to produce fine wool in sheep, are improvements which have been made in particular varieties in the species to which these animals belong. What has been produced by art must be continued by the same means, for the most improved breeds will soon return to a state of nature, or perhaps defects will arise which did not exist when the breed was in its natural state, unless the greatest attention is paid to the selection of the individuals who are to breed together."

The exact origin of the common domestic fowl and its numerous varieties is unknown. It is doubtless derived from one or more of the wild or jungle fowls of India. Some naturalists are of opinion that it is derived from the common jungle fowl known as the *Gallus Bankiva* of Temminck, or *Gallus Ferrugineus* of Gmelin, which very closely resembles the variety known as Black-breasted Red Game, except that the tail of the cock is more depressed; while others consider it to have been produced by the crossing of that species with one or more others, as the Malay gigantic fowl, known as the *Gallus Giganteus* of Temminck, Sonnerat's Jungle Fowl, *Gallus Sonneratii*, and probably some other species. At what period or by what people it was reclaimed is not known, but it was probably first domesticated in India. The writers of antiquity speak of it as a bird long domesticated and widely spread in their days. Very likely there are many species unknown to us in Sumatra, Java, and the rich woods of Borneo.

The process by which the various breeds have been produced "is simple and easily understood," says Mr. Wright. "Even in the wild state the original breed will show *some* amount of variation in colour, form, and size; whilst in domestication the tendency to change, as every one knows, is very much increased. By breeding from birds which show any marked feature, stock is obtained of which a portion will possess that feature in an *increased degree*; and by again selecting the best specimens, the special points sought may be developed to almost any degree required. A good example of such a process of development may be seen in the 'white face' so conspicuous in the Spanish breed. White ears will be observed occasionally in all fowls; even in such breeds as Cochins or Brahmas, where white ear-lobes are considered almost fatal blemishes; they continually occur, and by selecting only white-eared specimens to breed from, they might be speedily fixed in any variety as one of the characteristics. A large pendent white ear-lobe once firmly established, traces of the white *face* will now and then be found, and by a similar method is capable of development and fixture; whilst any colour of plumage or of leg may be obtained and established in the same way. The original amount of character required is *very* slight; a single hen-tailed cock will be enough to give that characteristic to a whole breed. Any peculiarity of *constitution*, such as constant laying, or frequent incubation, may be developed and perpetuated in a similar manner, all that is necessary being care and time. That such has been the method employed in the formation of the more distinct races of our poultry, is proved by the fact that a continuance of the same careful selection is needful to perpetuate them in perfection. If the very best examples of a breed are selected as the starting-point, and the produce is bred from indiscriminately for many generations, the distinctive points, whatever they are, rapidly decline, and there is also a more or less gradual but sure return to the primitive wild type, in size and even colour of the plumage. The purest black or white originally, rapidly becomes first marked with, and ultimately changed into, the original red or brown, whilst the other features simultaneously disappear. If, however, the process of artificial selection be carried too far, and with reference *only to one* prominent point, any breed is almost sure to suffer in the other qualities which have been neglected, and this has been the case with the very breed already mentioned—the white-faced Spanish. We know from old fanciers that this breed was formerly considered hardy, and even in winter rarely failed to afford a constant supply of its unequalled large white eggs. But of late years attention has been so *exclusively* directed to the 'white-face,' that whilst this feature has been developed and perfected to a degree never before known, the breed has become one of the most delicate of all, and the laying qualities of at least many strains have greatly fallen off. It would be difficult to avoid such evil results if it were not for a valuable compensating principle, which admits of *crossing*. That principle is, that any desired point possessed in perfection by a foreign breed, may be introduced by crossing into a strain it is desired to improve, and every other characteristic of the cross be, by selection, afterwards bred out again. Or one or more of these additional characteristics may be also retained, and thus a *new variety* be established, as many have been within the last few years."

Size may be imparted to the Dorking by crossing it with the Cochin, and the disposition to feather on the legs bred out again by judicious selection ; and the constitution may be strengthened by crossing with the Game breed. Game fowls that have deteriorated in size, strength, and fierceness, by a long course of breeding in-and-in, may have all these qualities restored by crossing with the fierce, powerful, and gigantic Malay, and his peculiarities may be afterwards bred out. The size of the eggs of the Hamburg might very probably be increased without decreasing, or with very slightly decreasing, the number of eggs, by crossing with a Houdan cock ; and the size would also be increased for the table. The French breeds, Crêve Cœur, Houdan, and La Flêche, gain in size and hardiness by being crossed with the Brahma cock. The cross between a Houdan cock and a Brahma hen " produces," says the " Henwife," " the finest possible chickens for market, but not to breed from. Pure Brahmas and Houdans alone must be kept for that purpose; I have always found the second cross worthless."

In crossing, the cockerels will more or less resemble the male, and the pullets the hen. "Long experience," says Mr. Wright, " has ascertained that the male bird has most influence upon the colour of the progeny, and also upon the comb, and what may be called the 'fancy points,' of any breed generally ; whilst the form, size, and useful qualities are principally derived from the hen."

Breed only from the strongest and healthiest fowls. In the breeding of poultry it is a rule, as in all other cases of organised life, that the best-shaped be used for the purpose of propagation. If a cock and hen have both the same defect, however trifling it may be, they should never be allowed to breed together, for the object is to improve the breed, not to deteriorate it, even in the slightest degree.

G

Hens should never be allowed to associate with a cock of a different breed if you wish to keep the breed pure, and if you desire superior birds, not even with an inferior male of their own variety. "No time," says Mr. Baily, "has ever been fixed as necessary to elapse before hens that have been running with cocks of divers breeds, and afterwards been placed with their legitimate partners, can be depended upon to produce purely-bred chickens; I am disposed to think at least two months. Time of year may have much to do with it. In the winter the escape of a hen from one run to the other, or the intrusion of a cock, is of little moment; but it may be serious in the spring, and destroy the hopes of a season." Many poultry-keepers separate the cocks and hens after the breeding season, considering that stronger chickens will be thereby obtained the next season. Where there is a separate house and run for the sitting hen this can be conveniently done when that compartment is vacant. In order to preserve a breed perfectly pure, it will be necessary, where there is not a large stock of the race, to breed from birds sprung from the same parents, but the blood should be crossed every year by procuring one or more fowls of the same breed from a distance, or by the exchange of eggs with some neighbouring stock, of colour and qualities as nearly allied as possible with the original breed.

CHAPTER IX.

POULTRY SHOWS.

A FEW years ago poultry shows were unknown. In 1846, the first was held in the Gardens of the Zoological Society, in the Regent's Park; Mr. Baily being the sole judge. It was a very fair beginning, but did not succeed, and it was not till the Cochin-China breed was introduced into this country, and the first Birmingham show was held, that these exhibitions became successful.

In 1849, "the first poultry show that was ever held in 'the good old town of Birmingham,' was beset with all the untried difficulties of such a scheme, when without the experiences of the present day, then altogether unavailable, a few spirited individuals carried to a successful issue an event that has now proved the foster-parent of the many others of similar character that abound in almost every principal town of the United Kingdom. It is quite essential, that I may be clearly understood, to preface my narrative by assuring fanciers that in those former days poultry amateurs were by no means as general as at the present time; few and far between were their locations; and though even then, among the few who felt interest in fowls, emulation existed, generally speaking, the keeping of poultry was regarded as 'a useless hobby,' 'a mere individual caprice,' 'an idle whim from which no good result could by possibility accrue'; nay, sometimes it was hinted, 'What a pity they have not something better to employ them during leisure hours!' and they were styled 'enthusiasts.' But have not the records of every age proved that enthusiasts are invariably the pioneers of improvement? And time, too, substantiated the verity of this rule in reference to our subject; for, among other proofs, it brought incontestable evidence that the raising of poultry was by no means the unremunerative folly idlers supposed it to be, and hesitated not rashly to declaim it; likewise, that it simply required to be fairly brought under public notice, to prove its general utility, and to induce the acknowledgment of how strangely so important a source of emolument had been hitherto neglected and overlooked."

At the Birmingham Show of 1852, about five thousand fowls were exhibited, and the specimens sold during the four days of the show amounted to nearly two thousand pounds, notwithstanding the high prices affixed to the pens, and that many were placed at enormous prices amounting to a prohibition, the owners not wishing to sell them. The Birmingham shows now generally comprise from one to two thousand pens of fowls and water-fowls, arranged in nearly one hundred classes; besides an equal proportion of pigeons. This show is the finest and most important, but there are many others of very high character and great extent. Poultry is also now exhibited to a considerable extent at agricultural meetings.

Any one may see the wonderful improvement that has been made in poultry-breeding by visiting the next Birmingham or other first-class show, and comparing the fowls there exhibited with those of his earliest recollections, and with those mongrels and impure breeds which may still be seen in too many farmyards. Points that were said to be impossible of attainment have been obtained with comparative ease by perserverance and skill, and the worst birds of a show are now often superior to the chief prize fowls of former days. Indeed, "a modern prize bird," says the "Henwife," "almost merits the character which a Parisian waiter gave of a melon, when asked to pronounce whether it was a fruit or a vegetable, ' Gentlemen,' said he, ' a melon is neither; it is a work of art.' "

Such shows must have great influence on the improvement of the breeds and the general management of poultry, though like all other prize exhibitions they have certain disadvantages. "We cannot but think," says Mr. Wright, "that our poultry shows have, to some extent, by the character of the judging, hindered the improvement of many breeds.

It will be readily admitted in *theory* that a breed of fowls becomes more and more valuable as its capacity of producing eggs is increased, and the quantity and quality of its flesh are improved, with a small amount of bone and offal in proportion. But, if we except the Dorking, which certainly is judged to some extent as a table fowl, all this is *totally* lost sight of both by breeders and judges, and attention is fixed exclusively upon colour, comb, face, and other equally fancy 'points.' Beauty and utility might be *both* secured. The French have taught us a lesson of some value in this respect. Within a comparatively recent period they have produced, by crossing and selection, four new varieties, which, although inferior in some points to others of older standing, are all eminently valuable as table fowls; and which in one particular are superior to any English variety, not even excepting the Dorking—we mean the very small proportion of bone and offal. This is really useful and scientific breeding, brought to bear upon *one* definite object, and we do trust the result will prove suggestive with regard to others equally valuable. We should be afraid to say how much might be done if English breeders would bring *their* perseverance and experience to bear in a similar direction. Agricultural Societies in particular might be expected in *their* exhibitions to show some interest in the improvement of poultry regarded as *useful stock*, and to them especially we commend the matter."

The rules and regulations relating to exhibitions vary at different shows, and may be obtained by applying to the secretary. Notices of exhibitions are advertised in the local papers, and in the *Field* and other London papers of an agricultural character.

In breeding birds for exhibition the number of hens to one cock should not exceed four or five, but if only two or three hens of the breed are possessed, the proper number of his harem should be made up by the addition of hens of another breed, those being chosen whose eggs are easily known from the others.

If it is intended to rear the chickens for exhibition at

the June, July, or August shows, the earlier they are hatched the better, and therefore a sitting should be made in January, if you have a young, healthy hen broody. Set her on the ground in a warm, sheltered, and quiet place, perfectly secure from rain, or from any flow of snow water. Feed her well, and keep water and small quantities of food constantly within her reach, so that she may not be tempted to leave the nest in search of food; for the eggs soon chill in winter. Mix the best oatmeal with hot water, and give it to her warm twice a day. A few grains of hempseed as a stimulant may be given in the middle of the day. The great difficulty to overcome in rearing early chickens is to sustain their vital powers during the very long winter nights, when they are for so many hours without food, the only substitute for which is warmth, and this can only be well got from the hen. Consequently a young Cochin-China with plenty of "fluff" will provide most warmth. The hen should not be set on more than five, or at most seven eggs; for if she has more, although she may sufficiently cover the chickens while very small, she will not be able to do so when they grow larger, and the outer ones will be chilled unless they manage to push themselves into the inside places, and then the displaced chickens being warm are sure to get more chilled than the others; and so the greater number of the brood, even if they survive, will probably be weakly, puny things, through the greedy desire to rear so many, while if she hatch but five chickens she will probably rear four. The hen should be cooped until the chickens are at least ten weeks old, and covered up at night with matting, sacking, or a piece of carpet.

Give them plenty of curd, chopped egg, and oatmeal, mixed with new milk. Stiff oatmeal porridge is the best stock food. Some onion tops minced fine will be an excellent addition if they can be had. They should have some milk to drink. Feed the hen well. The best warmth the chickens can have is that of their mother, and the best warmth for her is generated by generous, but proper, food, and a good supply of it. Early chickens rearing for show should be fed twice after dark, say at eight and eleven o'clock, and again at seven in the morning, so that they will not be without food for more than eight hours. The hen should be fed at the same times, and she will become accustomed to it, and call the chickens to feed; it will also generate more warmth in her for their benefit. Yolk of egg beaten up and given to drink is most strengthening for weakly chickens; or it may be mixed with their oatmeal. The tender breeds should not be hatched till April or May, unless in a mild climate, or with exceptional advantages.

For winter exhibition, March and April hatched birds are preferable to those hatched earlier. Not more than seven eggs should be set, for a hen cannot scratch up insects and worms and find peculiar herbage for more than six chickens. If the chickens have not a good grass run, they must be supplied with abundance of green food.

They should not be allowed to roost before they are three months old, and the perches must be sufficiently large. Mr. Wright recommends a bed of clean, dry ashes, an inch deep, for those that leave the hen before the proper age for roosting, and does not allow his chickens, even while with the hen, to bed upon straw, considering the ashes to be much cleaner and also warmer.

The chickens intended to be exhibited should be distinguished from their companions by small stripes of different coloured silks loosely sewn round their legs, which distinguishing colours should be entered in the poultry-book. A few good birds should always be kept in reserve to fill up the pen in case of accidents.

Weight is more important in the December and later winter shows than at those held between August and November, but at all shows feather and other points of competitors being equal weight must carry the day, Game and Bantams excepted. It is not safe to trust to the apparent weight of a bird, for the feathers deceive, and it is therefore advisable to weigh the birds occasionally. Each should be weighed in a basket, allowance being made for the weight of the basket, and they should if possible be weighed before a meal. But fowls that are over-fattened, as some judges very improperly desire, cannot be in good health any more than "crammed" fowls, and are useless for breeding, producing at best a few puny, delicate, or sickly chickens; thus making the exhibition a mere "show," barren of all useful results.

Pullets continue to grow until they begin to lay, which almost or quite stops their growth; and therefore if great size is desired for exhibition, they should be kept from the cockerels and partly from stimulating food until a month before the show, when they will be required to be matched in pens. During this month they should have extra food and attention.

If fowls intended for exhibition are allowed to sit, the chickens are apt to cause injury to their plumage, and loss of condition, while if prevented from sitting, they are liable to suffer in moulting. Their chickens may be given to other hens, but the best and safest plan is to set a broody exhibition hen on duck's eggs, which will satisfy her natural desire for sitting, while the young ducklings will give her much less trouble, and leave her sooner than a brood of her own kind.

All the birds in a pen should match in comb, colour of their legs, and indeed in every particular. Mr. Baily mentions "a common fault in exhibitors who send two pens composed of three excellent and three inferior birds, so divided as to form perhaps one third class and one highly commended pen: whereas a different selection would make one of unusual merit. If an amateur who wishes to exhibit has fifteen fowls to choose from, and to form a pen of a cock and two hens, he should study and scan them closely while feeding at his feet in the morning. He should then have a place similar to an exhibition pen, wherein he can put the selected birds; they should be raised to the height at which he can best see them, and before he has looked long at them defects will become apparent one after the other till, in all probability, neither of the subjects of his first selection will go to the show. We also advise him rather to look for defects than to dwell on beauties—the latter are always prominent enough. The pen of which we speak should be a moveable one for convenience' sake, and it is well to leave the fowls in it for a time that they may become accustomed to each other, and also to an exhibition pen." Birds that are strangers should never be put into the same hamper, for not only the cocks but even the hens will fight with and disfigure each other.

Some give linseed for a few days before the exhibition to impart lustre to the plumage, by increasing the secretion of oil. A small quantity of the meal should be mixed with their usual soft food, as fowls generally refuse the whole grain. But buckwheat and hempseed, mixed in equal proportions, if given for the evening meal during the last ten or twelve days, is healthier for the bird, much liked, and will not only impart equal lustre to the plumage, but also improve the appearance of the comb and wattles.

Spanish fowls should be kept in confinement for some days before the show, with just enough light to enable them to feed and perch, and the place should be littered with clean straw. This greatly improves their condition; why we know not, but it is an established fact. Game fowls should be kept in for a few days, and fed on meal, barley, and bread, with a few peas, which tend to make the plumage hard, but will make them too fat if given freely. Dark and golden birds should be allowed to run about till they have to be sent off. Remove all scurf or dead skin from the comb, dry dirt from the beak, and stains from the plumage, and wash their legs clean. White and light fowls that have a good grass run and plenty of clean straw in their houses and yards to scratch in, will seldom require washing, but town birds, and country ones if not perfectly clean, should be washed the day before the show with tepid water and mild white soap rubbed on flannel, care being taken to wash the feathers downwards, so as not to break or ruffle them; afterwards wiped with a piece of flannel that has been thoroughly soaked in clean water, and gently dried with soft towels before the fire; or the bird may be entirely dipped into a pan of warm water, then rinsed thoroughly in cold water, wiped with a flannel, and placed in a basket with soft straw before a fire to dry. They should then be shut up in their houses with plenty of clean straw. They should have their feet washed if dirty, and be well fed with soft nourishing food just before being put into the travelling-basket, for hard food is apt to cause fever and heat while travelling, and, having to be digested without gravel or exercise, causes indigestion, which ruffles the plumage, dulls its colour, darkens the comb, and altogether spoils the appearance of the bird. Sopped or steeped bread is excellent.

The hampers should always be round or oval in form, as fowls invariably creep into corners and destroy their plumage. They should be high enough for the cocks to stand upright in, without touching the top with their combs. Some exhibitors prefer canvas tops to wicker lids, considering that the former preserve the fowls' combs from injury if they should strike against the top, while others prefer the latter as being more secure, and allowing one hamper to be placed upon another if necessary, and also preserving the fowls from injury if a heavy hamper or package should otherwise be placed over it. A good plan is to have a double canvas top, the space between being filled with hay. A thick layer of hay or straw should be placed at the bottom of the basket. Wheaten straw is the best in summer and early autumn, and oat or barley straw later in the year and during winter. A good lining also is essential; coarse calico stitched round the inside of the basket is the best. Ducks and geese do not require their hampers to be lined, except in very cold weather; and the best lining for them is made by stitching layers of pulled straw round the inside of the basket. Turkeys should have their hampers lined, for although they are very hardy, cold and wet damage their appearance more than other poultry. Take care that the geese cannot get at the label, for they will eat it, and also devour the hempen fastenings if within their reach.

Be very careful in entering your birds for exhibition;

describe their ages, breed, &c., exactly and accurately, and see yourself to the packing and labelling of their hampers.

Mr. F. Wragg, the superintendent of the poultry-yard of R. W. Boyle, Esq., whose fowls have a sea voyage from Ireland besides the railway journey, and yet always appear in splendid condition and "bloom," ties on one side of the hamper, "near the top, a fresh-pulled cabbage, and on the other side a good piece of the bottom side of a loaf, of which they will eat away all the soft part. Before starting, I give each bird half a tablespoonful of port wine, which makes them sleep a good part of the journey. Of course, if I go with my birds, as I generally do, I see that they, as well as myself, have 'refreshment' on the road." * The cabbage will always be a treat, and the loaf and wine may be added for long journeys.

Birds are frequently overfed at the show, particularly with barley, which cannot be properly digested for want of gravel and exercise; and therefore, if upon their return their crops are hard and combs look dark, give a tablespoonful of castor oil; but if they look well do not interfere with them. They should not have any grain, but be fed sparingly on stale bread soaked in warm ale, with two or three mouthfuls of tepid water, for liquid is most hurtful if given in quantity. They should not be put into the yard with the other fowls which may treat them, after their absence, as intruders, but be joined with them at night when the others have gone to roost. On the next day give them a moderate allowance of soft food with a moderate supply of water, or stale bread sopped in water, and a sod of grass or half a cabbage leaf each, but no other green food; and on the following day they may have their usual food.

When the fowls are brought back, take out the linings, wash them, and put them by to be ready for the next show; and after the exhibition season, on a fine dry day, wash the hampers, dry them thoroughly, and put them in a dry place. Never use them as quiet berths for sick birds, which are sure to infect them and cause the illness of the next occupants; or as nesting-places for sitting hens, which may leave insects in the crevices that will be difficult to eradicate.

* The Practical Poultry Keeper. By Mr. L. Wright. Cassell, Petter & Galpin.

In our descriptions of the various Breeds, we have given sufficient general information upon the Exhibition Points from the best authorities; but considerable differences of opinion have been expressed of late years, and eminent breeders dissent in some cases even from the generally recognised authority of the popular "Standard of Excellence." We, therefore, advise intending exhibitors to ascertain the standards to be followed at the show and the predilections of the judges, and to breed accordingly, or, if they object to the views held, not to compete at that exhibition.

TECHNICAL TERMS.

Coverts.—The *upper* and *lower wing coverts* are those ranges of feathers which cover the primary quills; and the *tail coverts* are those feathers growing on each side of the tail, and are longer than the body feathers, but shorter than those of the tail.

Dubbing.—Cutting off the comb and wattles of a cock; an operation usually confined to Game cocks.

Ear-lobe.—The small feathers covering the organ of hearing, which is placed a little behind the eye.

Flight.—The last five feathers of each wing.

Fluff.—The silky feathers growing on the thighs and hinder parts of Cochin-China fowls.

Hackles.—The *neck hackles* are feathers growing from the neck, and covering the shoulders and part of the back; and the *saddle hackles* those growing from the end of the back, and falling over the sides.

Legs.—The *legs* are properly the lower and scaly limbs, the upper part covered with feathers and frequently miscalled legs, being correctly styled the *thighs*.

Primary Quills.—The long, strong quills, usually ten in number, forming the chief portion of each wing, and the means of flight.

Vulture-hocked.—Feathers growing from the thigh, and projecting backwards below the knee.

Buff and White Cochin-China.   Malay Cock.   Light and Dark Brahmahs.

CHAPTER X.

COCHIN-CHINAS, OR SHANGHAES.

LIKE many other fowls these possess a name which is incorrectly applied, for they came from Shanghae, not Cochin-China, where they were comparatively unknown. Mr. Fortune, who, from his travels in China, is well qualified to give an opinion, states that they are a Chinese breed, kept in great numbers at Shanghae; the real Cochin-China breed being small and elegantly shaped. But all attempts to give them the name of the port from which they were brought have failed, and the majority of breeders persist in calling them Cochins. In the United States both names are used, the feather-legged being called Shanghaes, and the clean-legged Cochins.

The first Shanghae fowls brought to this country were sent from India to Her Majesty, which gave them great importance; and the eggs having been freely distributed by the kindness of the Queen and the Prince Consort, the breed was soon widely spread. They were first introduced into this country when the northern ports of China, including Shanghae, were thrown open to European vessels on the conclusion of the Chinese war in 1843; but some assign the date of their introduction from 1844 to 1847, and say that those called Cochins, exhibited by the Queen in 1843, were not the true breed, having been not only entirely without feathers on the shanks, but also altogether different in form and general characteristics. A pair which were sent by Her Majesty for exhibition at the Dublin Cattle Show in April, 1846, created such a sensation from their great size and immense weight, and the full, loud, deep-pitched crowing of the cock, that almost every one seemed desirous to possess some of the breed, and enormous prices were given for the eggs and chickens.

With his propensity for exaggeration, Paddy boasted that they laid five eggs in two days, each weighing three ounces, that the fowls equalled turkeys in size, and "Cochin eggs became in as great demand as though they had been laid by the fabled golden goose. Philosophers, poets, merchants, and sweeps had alike partook of the mania; and although the latter could hardly come up to the price of a real Cochin, there were plenty of vagabond dealers about, with counterfeit crossed birds of all kinds, which were advertised to be the genuine article. For to such a pitch did the excitement rise, that they who never kept a fowl in their lives, and would hardly know a Bantam from a Dorking, puzzled their shallow brains as to the proper place to keep them, and the proper diet to feed them on." Their justly-deserved popularity speedily grew into a mania, and the price which had been from fifteen to thirty shillings each, then considered a high price for a fowl, rose to ten pounds for a fine specimen, and ultimately a hundred guineas was repeatedly paid for a single cock, and was not an uncommon price for a pair of really fine birds. "They were afterwards bred," says Miss Watts, "for qualities difficult of attainment, and, as the result proved, little worth trying for," and "fowls with *many* excellent qualities were blamed for not being *perfect*," and they fell from their high place, and were as unjustly depreciated as they had been unduly exalted.

"Had these birds," wrote Mr. Baily many years since, "been shy breeders—if like song birds the produce of a pair were four, or at most five, birds in the year, prices might have been maintained; but as they are marvellous layers they increased. They bred in large numbers, and consequently became cheaper, and then the mania ended, because those who dealt most largely in them did so not from a love of the birds or the pursuit, but as a speculation. As they had over-praised them before, they now treated them with contempt. Anything like a moderate profit was despised, and the birds were left to their own merits. These were sufficient to ensure their popularity, and now after fluctuating in value more than anything except shares, after being over-praised and then abused, they have remained favourites with a large portion of the public, sell at a remunerating price, and form one of the largest classes at all the great exhibitions." This has proved to be a perfectly correct view, and the breed is now firmly established in public estimation, and unusually fine birds will still sell for from five to twenty pounds each. The mania did great service to the breeding and improvement of poultry by awakening an interest in the subject throughout the kingdom which has lasted.

They are the best of all fowls for a limited space, and not inclined to wander even when they have an extensive run. They cannot fly, and a fence three feet high will keep them in. But if kept in a confined space they must have an unlimited supply of green food. They give us eggs when they are most expensive, and indeed, with regard to new-laid eggs, when they are almost impossible to be had at any price. They begin to lay soon after they are five months old, regardless of the season or weather, and lay throughout the year, except when requiring to sit, which they do twice or thrice a year, and some oftener. Pullets will sometimes lay at fourteen weeks, and want to sit before they are six months old. Cochins have been known to lay twice in a day, but not again on the following day, and the instances are exceptional. Their eggs are of a pale chocolate colour, of excellent flavour, and usually weigh $2\frac{1}{4}$ ounces each. They are excellent sitters and mothers. Pullets will frequently hatch, lay again, and sit with the chickens of the first brood around them. Cochins are most valuable as sitters early in the year, being broody when other fowls are beginning to lay; but unless cooped they are apt to leave their chickens too soon, especially for early broods, and lay again. They are very hardy, and their chickens easy to rear, doing well even in bleak places without any unusual care. But they are backward in fledging, chickens bred from immature fowls being the most backward. Those which are cockerels show their flight feathers earliest. They are very early matured.

A writer in the *Poultry Chronicle* well says: "These

fowls were sent to provide food for man; by many they are not thought good table fowls; but when others fail, if you keep them, you shall never want the luxury of a really new-laid egg on your breakfast table. The snow may fall, the frost may be thick on your windows when you first look out on a December morning, but your Cochins will provide you eggs. Your children shall learn gentleness and kindness from them, for they are kind and gentle, and you shall be at peace with your neighbours, for they will not wander nor become depredators. They have fallen in price because they were unnaturally exalted; but their sun is not eclipsed; they have good qualities, and valuable. They shall now be within the reach of all; and will make the delight of many by their domestic habits, which will allow them to be kept where others would be an annoyance." They will let you take them off their roost, handle and examine them, and put them back without struggling.

The fault of the Cochin-Chinas as table birds is, that they produce most meat on the inferior parts; thus, there is generally too little on the breast which is the prime part of a fowl, while the leg which is an inferior part, is unusually fleshy, but it must be admitted that the leg is more tender than in other breeds. A greater quantity of flesh may be raised within a given time, on a certain quantity of food, from these fowls than from any other breed. The cross with the Dorking is easily reared, and produces a very heavy and well-shaped fowl for the table, and a good layer.

"A great hue and cry," says Miss Watts, "has been raised against the Cochin-Chinas as fowls for the table, but we believe none have bestowed attention on breeding them with a view to this valuable consideration. Square, compact, short-legged birds have been neglected for a certain colour of feather, and a broad chest was given up for the wedge-form at the very time that was pronounced a fault in the fowl. It is said that yellow-legged fowls are yellow also in the skin, and that white skin and white legs accompany each other; but how pertinaciously the yellow leg of the Cochin is adhered to! Yet all who have bred them will attest that a little careful breeding would perpetuate white-legged Cochins. Exhibitions are generally excellent; but to this fowl they certainly have only been injurious, by exaggerating useless and fancy qualities at the expense of those which are solid and useful. Who would favour, or even sanction, a Dorking in which size and shape, and every property we value in them, was sacrificed to an endeavour to breed to a particular colour? and this is what we have been doing with the Cochin-China. Many breeders say, eat Cochins while very young; but we have found them much better for the table as fowls than as chickens. A fine Cochin, from five to seven months old, is like a turkey, and very juicy and fine in flavour."

A peculiar characteristic of these birds, technically called "fluff," is a quantity of beautifully soft, long feathers, covering the thighs till they project considerably, and garnishing all the hinder parts of the bird in the same manner, so that the broadest part of the bird is behind. Its quality is a good indication of the breed; if fine and downy the birds are probably well-bred, but if rank and coarse they are inferior. The cocks are frequently somewhat scanty in "fluff," but should be chosen with as much as possible; but vulture-hocks which often accompany the heaviest feathered birds should be avoided, as they now disqualify at the best shows. "The fluff," says a good authority, "in the hen especially, should so cover the tail feathers as to give the appearance of a very short back, the line taking an upward direction from within an inch or so of the point of junction with the hackle." The last joint of the wings folds up, so that the ends of the flight feathers are concealed by the middle feathers, and their extremities are again covered by the copious saddle, which peculiarity has caused them to be also called the ostrich-fowl.

A good Cochin cock should be compact, large, and square built; broad across the loins and hind-quarters; with a deep keel; broad, short back; short neck; small, delicately-shaped, well-arched head; short, strong, curved beak; rather small, finely and evenly serrated, straight, single, erect comb, wholly free from reduplications and sprigs; brilliant red face, and pendant wattles; long hanging ear-lobe, of pure red, white being inadmissible; bright, bold eye, approaching the plumage in colour; rich, full, long hackle; small, closely-folded wings; short tail, scarcely any in some fine specimens, not very erect, with slightly twisted glossy feathers falling over it like those of the ostrich; stout legs set widely apart, yellow and heavily feathered to the toe; and erect carriage. The chief defect of the breed is narrowness of breast, which should therefore be sought for as full as possible.

The hen's body is much deeper in proportion than that of the cock. She resembles him upon most points, but differs in some; her comb having many indentations; the fluff being softer, and of almost silky quality; the tail has upright instead of falling feathers, and comes to a blunt point; and her carriage is less upright.

Cochins lose their beauty earlier than any other breed, and moult with more difficulty each time. They are in their greatest beauty at from nine to eighteen months old. The cocks' tails increase with age. In buying Cochins avoid clean legs, fifth toes, which show that it has been crossed with the Dorking, double combs that betray Malay blood, and long tails, particularly taking care that the cock has not, and ascertaining that he never had, sickle feathers. The cock ought not to weigh less than ten or eleven pounds, and a very fine bird will reach thirteen; the hens from eight to ten pounds.

The principal colours now bred are Buff, Cinnamon, Partridge, Grouse, Black, and White. The Buff and White are the most popular.

Buff birds may have black in the tails of both sexes, but the less there is the better. Black-pencilling in the hackle is considered objectionable at good shows. The cock's neck hackles, wing coverts, back, and saddle hackles, are usually of a rich gold colour, but his breast and the lower parts of his body should match with those of his hens. Buff birds generally produce chickens lighter than themselves. Most birds become rather lighter at each moult. In making up an exhibition pen, observe that

Grouse and Partridge hens should have a black-breasted cock; and that Buff and Cinnamon birds should not be placed together, but all the birds in the pen should be either Buff or Cinnamon. The Cinnamon are of two shades, the Light Cinnamon and the Silver, which is a pale washy tint, that looks very delicate and pretty when perfectly clean. Silver Cinnamon hens should not be penned with a pale Yellow cock, but with one as near to their own tint as can be found. Mr. Andrews's celebrated strain of Cochins sometimes produced both cocks and hens which were Silver Cinnamon, with streaks of gold in the hackle.

In Partridge birds the cock's neck and saddle hackles should be of a bright red, striped with black, his back and wings of dark red, the latter crossed with a well-defined bar of metallic greenish black, and the breast and under parts of his body should be black, and not mottled. The hen's neck hackles should be of bright gold, striped with black, and all the other portions of her body of a light brown, pencilled with very dark brown. The Grouse are very dark Partridge, have a very rich appearance, and are particularly beautiful when laced. They are far from common, and well worth cultivating. The Partridge are more mossed in their markings, and not so rich in colour as the Grouse. Cuckoo Cochins are marked like the Cuckoo Dorkings, and difficult to breed free of yellow.

The White and Black were introduced later than the others. Mr. Baily says the White were principally bred from a pair imported and given to the Dean of Worcester, and which afterwards became the property of Mrs. Herbert, of Powick. White Cochins for exhibition must have yellow legs, and they are prone to green. The origin of the Black is disputed. It is said to be a sport from the White, or to have been produced by a cross between the Buff and the White. By careful breeding it has been fixed as a decided sub-variety, but it is difficult, if not almost impossible, to rear a cock to complete maturity entirely free from coloured feathers. They keep perfectly pure in colour till six months old, after which age they sometimes show a golden patch or red feathers upon the wing, or a few streaks of red upon the hackle, of so dark a shade as to be imperceptible except in a strong light, and are often found on close examination to have white under feathers, and others barred with white.

The legs in all the colours should be yellow. Flesh-coloured legs are admissible, but green, black, or white are defects. In the Partridge and Grouse a slight wash, as of indigo, appears to be thrown over them, which in the Black assumes a still darker shade; but in all three yellow should appear partially even here beneath the scales, as the pink tinge does in the Buff and White birds.

Cochin-Chinas being much inclined to accumulate internal fat, which frequently results in apoplexy, should not be fed on food of a very fattening character, such as Indian corn. They are liable to have inflamed feet if they are obliged to roost on very high, small, or sharp perches, or allowed to run over sharp-edged stones.

They are also subject to an affection called White Comb, which is a white mouldy eruption on the comb and wattles like powdered chalk; and if not properly treated in time, will spread over the whole body, causing the feathers to fall off. It is caused by want of cleanliness, over-stimulating or bad food, and most frequently by want of green food, which must be supplied, and the place rubbed with an ointment composed of two parts of cocoanut oil, and one of turmeric powder, to which some persons add one half part of sulphur; and six grains of jalap may be given to clear the bowels.

CHAPTER XI.

BRAHMA-POOTRAS.

It is a disputed point among great authorities whether Brahmas form a distinct variety, or whether they originated in a cross with the Cochin, and have become established by careful breeding. When they were first introduced, Mr. Baily considered them to be a distinct breed, and has since seen nothing to alter his opinion. Their nature and habits are quite dissimilar, for they wander from home and will get their own living where a Cochin would starve, have more spirit, deeper breasts, are hardier, lay larger eggs, are less prone to sit, and never produce a clean-legged chicken. Whatever their origin, by slow and sure degrees, without any mania, they have become more and more popular, standing upon their own merits, and are now one of the most favourite varieties.

"The worst accusation," says Miss Watts, "their enemies can advance against them is, that no one knows their origin; but this is applicable to them only as it is when applied to Dorkings, Spanish, Polands, and all the other kinds which have been brought to perfection by careful breeding, working on good originals. All we have in England are descended from fowls imported from the United States, and the best account of them is, that a sailor (rather vague, certainly) appeared in an American town (Boston or New York, I forget which) with a new kind of fowl for sale, and that a pair bought from him were the parents of all the Brahmas. Uncertain as this appears, the accounts of those who pretend to trace their origin as cross-bred fowls is, at least, equally so, and I believe we may just act towards the Brahmas as we do with regard to Dorkings and other good fowls, and be satisfied to possess a first-rate, useful kind, although we may be unable to trace its genealogical tree back to the root. Whatever may be their origin, I find them distinct in their characteristics. I have found them true to their points, generation after generation, in all the years that I have kept them. The pea-comb is very peculiar, and I have never had one chicken untrue in this among all that I have bred. Their habits are very unlike the Cochins. Although docile, they are much less inert; they lay a larger number of eggs, and sit less frequently. Many of my hens only wish to sit once a year; a few oftener than that, perhaps twice or even three times in rare instances, but never at the end of each small batch of eggs, as I find (my almost equal favourites) the Cochins do. The division of Light and Dark Brahmas is a fancy of the judges, which any one who keeps them can humour with a little care in breeding. My idea of their colour is, that it should be black and grey (iron grey, with more or less of a blue tinge, and devoid of any brown) on a clear white ground, and I do not care whether the white or the marking predominates. I believe breeders could bear me out, if they would, when I say many fowls which pass muster as Brahmas are the result of a cross, employed to increase size and procure the heavy colour which some of the judges affect."

For strength of constitution, both as chickens and fowls, they surpass all other breeds. Brahmas like an extensive range, but bear confinement as well as any fowls, and keep cleaner in dirty or smoky places than any that have white feathers. They are capital foragers where they have their liberty, are smaller eaters and less expensive to keep than Cochins, and most prolific in eggs. They lay regularly on an average five fine large eggs a week all the year round, even when snow is on the ground, except when moulting or tending their brood. Mr. Boyle, of Bray, Ireland, the most eminent breeder of Dark Brahmas in Great Britain, says he has "repeatedly known pullets begin to lay in autumn, and *never stop*— let it be hail, rain, snow, or storm— for a single day till next spring." They usually lay from thirty to forty eggs before they seek to sit. The hens do not sit so often as Cochins, and a week's change of place will generally banish the desire. They put on flesh well, with plenty of breast-meat, and are more juicy and better shaped for the table than most Cochins; though, after they are six months old, the flesh is much inferior to that of the Dorking. A cross with a Dorking or Crêve Cœur cock produces the finest possible table fowl, carrying almost incredible quantities of meat of excellent quality.

The chickens are hardy and easy to rear. They vary in colour when first hatched, being all shades of brown, yellow, and grey, and are often streaked on the back and spotted about the head; but this variety gives place, as the feathers come, to the mixture of black, white, and grey, which forms the distinguishing colour of the Brahma. Mr. Baily has "hatched them in snow, and reared them all out of doors without any other shelter than a piece of mat or carpet thrown over the coop at night." They reach their full size at an early age, and the pullets are in their prime at eight months. Miss Watts noticed that Brahmas "are more clever in the treatment of themselves when they are ill than other fowls; when they get out of order, they will generally fast until eating is no longer injurious," which peculiarity is corroborated by the experienced "Henwife." The feathers of the Brahma-Pootra are said to be nearly equal to goose feathers.

The head should have a slight fulness over the eye, giving breadth to the top; a full, pearl eye is much admired, but far from common; comb either a small single, or pea-comb—the single resembling that of the Cochin; the neck short; the breast wide and full; the legs short, yellow, and well-feathered, but not so fully as in the finest Cochins; and the tail short but full, and in the cock opening into a fan. They should be wide and deep made, large and weighty, and have a free, noble carriage, equally distinct from the waddle of the Cochin, and the erect bearing of the Malay. Unlike the Cochins, they keep constantly to their colour, which is a mixture of black, white, and grey; the lightest being almost white, and the darkest consisting of grey markings on a white ground.

The colour is entirely a matter of taste, but the bottom colour should always be grey.

"After breeding Brahmas for many years," says Miss Watts, "through many generations and crosses (always, however, keeping to families imported direct from America), we are quite confirmed in the opinion that the pea-comb is *the* comb for the Brahma; and this seems now a settled question, for single-combed birds never take prizes when passable pea-combed birds are present. The leading characteristic of the peculiar comb, named by the Americans the pea-comb, is its triple character. It may be developed and separated almost like three combs, or nearly united into one; but its triple form is always evident. What we think most beautiful is, where the centre division is a little fluted, slightly serrated, and flanked by two little side combs. The degree of the division into three varies, and the peculiarities of the comb may be less perceptible in December than when the hens are laying; but the triple character of the pea-comb is always evident. It shows itself in the chick at a few days old, in three tiny paralleled lines." It is thick at the base, and like three combs joined into one, the centre comb being higher than the other, but the comb altogether must be low, rounded at the top, and the indentations must not be deep. Whether single or triple, all the combs in a pen should be uniform.

The dark and light varieties should not be crossed, as, according to Mr. Teebay, who was formerly the most extensive and successful breeder of Brahmas in England, the result is never satisfactory.

CHAPTER XII.

MALAYS.

This was the first of the gigantic Asiatic breeds imported into this country, and in height and size exceeds any fowl yet known. The origin of the Malay breed is supposed to be the *Gallus giganteus* of Temminck. "This large and very remarkable species," says Mr. W. C. L. Martin, "is a native of Java and Sumatra. The comb is thick and low, and destitute of serrations, appearing as if it had been partially cut off; the wattles are small, and the throat is bare. The neck is covered with elongated feathers, or hackles, of a pale golden-reddish colour, which advance upon the back, and hackles of the same colour cover the rump, and drop on each side of the base of the tail. The middle of the back and the shoulders of the wings are of a dark chestnut, the feathers being of a loose texture. The greater wing-coverts are of a glossy green, and form a bar of that colour across the wing. The primary and secondary quill feathers are yellowish, with a tinge of rufous. The tail feathers are of a glossy green. The under surface uniformly is of a glossy blackish green, but the base of each feather is a chestnut, and this colour appears on the least derangement of the plumage. The limbs are remarkably stout, and the robust tarsi are of a yellow colour. The voice is a sort of crow—hoarse and short, and very different from the clear notes of defiance uttered by our farmyard chanticleer. This species has the habit, when fatigued, of resting on the tarsi or legs, as we have seen the emu do under similar circumstances."

In the "Proceedings of the Zoological Society" for 1832, we find the following notice respecting this breed, by Colonel Sykes, who observed it domesticated in the Deccan: "Known by the name of the Kulm cock by

Europeans in India. Met with only as a domestic bird; and Colonel Sykes has reason to believe that it is not a native of India, but has been introduced by the Mussulmans from Sumatra or Java. The iris of the real game bird should be whitish or straw yellow. Colonel Sykes landed two cocks and a hen in England in June, 1831. They bore the winter well; the hen laid freely, and has reared two broods of chickens. The cock has not the shrill clear pipe of the domestic bird, and his scale of note appears more limited. A cock in the possession of Colonel Sykes stood twenty-six inches high to the crown of the head; but they attain a greater height. Length from the tip of the bill to the insertion of the tail, twenty-three inches. Hen one-third smaller than the male. Shaw very justly describes the habit of the cock, of resting, when tired, on the first joint of the leg."

It is a long, large, heavy bird, standing remarkably upright, having an almost uninterrupted slope from the head to the insertion of the tail; with very long, though strong, yellow legs, quite free from feathers; long, stout, firm thighs, and stands very erect; the cock, when full grown, being at least two feet six inches, and sometimes over three feet in height, and weighing from eight to eleven pounds. The head has great fulness over the eye, and is flattened above, resembling that of the snake. The small, thick, hard comb, scarcely rising from the head, and barely as long, like half a strawberry, resembles that of a Game fowl dubbed. The wattles are very small; the neck closely feathered, and like a rope, with a space for an inch below the beak bare of feathers. It has a hard, cruel expression of face; a brilliant bold eye, pearled around the edge of the lids; skinny red face; very strong curved yellow beak; and small, drooping tail, with very beautiful, though short, sickle feathers. The hen resembles the cock upon all these points, but is smaller.

Their colours now comprise different shades of red and deep chestnut, in combination with rich browns, and there are also black and white varieties, each of which should be uniform. The feathers should be hard and close, which causes it to be heavier than it appears.

Malays are inferior to most other breeds as layers, but the pullets commence laying early, and are often good winter layers. Their eggs, which weigh about $2\frac{1}{2}$ ounces each, are of a deep buff or pale chocolate colour, surpass all others in flavour, and are so rich that two of them are considered to be equal to three of ordinary fowls. They are nearly always fertile.

Their chief excellence is as table fowls, carrying, as they do, a great quantity of meat, which, when under a year old, is of very good quality and flavour. Crossed with the Spanish and Dorking, they produce excellent table fowls; the latter cross being also good layers.

Malays are good sitters and mothers, if they have roomy nests. Their chickens should not be hatched after June, as they feather slowly, and are delicate; but the adult birds are hardy enough, and seem especially adapted to crowded localities, such as courts and alleys. "Malays," says Mr. Baily, "will live anywhere; they will inhabit a back yard of small dimensions; they will scratch in the dust-hole, and roost under the water-butt; and yet not only lay well, but show in good condition when requisite." Like the Game fowl, it is terribly pugnacious, and in its native country is kept and trained for fighting. This propensity, which is still greater in confinement, is its greatest disadvantage. When closely confined they are apt to eat each other's feathers, the cure for which is turning them into a grass run, and giving them a good supply of lettuce leaves, with an occasional purgative of six grains of jalap. The Chittagong is said to be a variety of the Malay.

CHAPTER XIII.

GAME.

THIS is the kind expressly called the English breed by Buffon and the French writers, and is the noblest and most beautiful of all breeds, combining an admirable figure, brilliant plumage, and stately gait. It is most probably derived from the larger or continental Indian species of the Javanese, or Bankiva Jungle Fowl—the *Gallus Bankiva* of Temminck—which is a distinct species, distinguished chiefly from the Javanese fowl by its larger size. (*See* page 124.) Of this continental species, Sir W. Jardine states that he has seen three or four specimens, all of which came from India proper. The Game cock is the undisputed king of all poultry, and is unsurpassed for courage. The Malay is more cruel and ferocious, but has less real courage. Game fowls are in every respect fighting birds, and, although cock-fighting is now very properly prohibited by law, Game fowls are always judged mainly in reference to fighting qualities. But their pugnacious disposition renders them very troublesome, especially if they have not ample range, although it does not disqualify them for small runs to the extent generally supposed. A blow with his spur is dangerous, and instances have been recorded of very severe injuries inflicted upon children, even causing death. An old newspaper states that "Mr. Johnson, a farmer in the West Riding of Yorkshire, who has a famous breed of the Game fowl, has had the great misfortune to lose his little son, a boy of three years old, who was attacked by a Game cock, and so severely injured that he died shortly afterwards." High-bred hens are quite as pugnacious as the cocks. The chickens are very quarrelsome, and both cocks and hens fight so furiously, that frequently one-half of a brood is destroyed, and the other half have to be killed.

Game fowls are hardy when they can have liberty, but cannot be well kept in a confined space. They eat little, and are excellent for an unprotected place, because by their activity they avoid danger themselves, and by their courage defend their chickens from enemies. The hen is a prolific layer, and, if she has a good run, equal to any breed. The eggs, though of moderate size only, are remarkable for delicacy of flavour. She is an excellent sitter, and still more excellent mother. The chickens are easily reared, require little food, and are more robust in constitution than almost any other variety.

The flesh of the Game fowl is beautifully white, and superior to that of all other breeds for richness and delicacy of flavour. They should never be put up to fat, as they are impatient of confinement. "They are in no way fit for the fattening-coop," says Mr. Baily. "They cannot bear the extra food without excitement, and that is not favourable to obesity. Nevertheless, they have their merits. If they are reared like pheasants round a keeper's house, and allowed to run semi-wild in the woods, to frequent sunny banks and dry ditches, they will grow up like them; they will have little fat, but they will be full of meat. They must be eaten young; and a Game pullet four or five months old, caught up wild in this way, and killed two days before she is eaten, is, perhaps, the most delicious chicken there is in point of flavour."

The Game-fowl continues to breed for many years without showing any signs of decay, and in this respect is superior to the Cochin, Brahma, and even to the Dorking.

The cock's head should be long, but fine; beak long, curved, and strong; comb single, small, upright, and bright red; wattles and face bright red; eyes large and brilliant; neck long, arched, and strong; breast well developed; back short and broad between the shoulders, but tapering to the tail; thighs muscular, but short compared to the shanks; spur low; foot flat, with powerful claws, and his carriage erect. The form of the hen should resemble the above on a smaller scale, with small, fine comb and face, and wattles of a less intense red. The feathers of both should be very hard, firm, and close, very strong in the quills, and seem so united that it should be almost impossible to ruffle them, each feather if lifted up falling readily into its original place. Size is not a point of merit, from four to six pounds being considered sufficient, and better than heavier weights. Among the list of imperfections in Game cocks, Sketchley enumerates "flat sides, short legs, thin thighs, crooked or indented breast, short thin neck, imperfect eye, and duck or short feet."

"It is the custom," says Miss Watts, "consequently imperative, that all birds which are exhibited should have been dubbed, and this should not be done until the comb is so much developed that it will not spring again after the dubbing. This will be safe if the chicken is nearly six months old, but some are more set than others at a certain age. A keen pair of scissors is the best instrument with which to operate. Hold the fowl with a firm hand, cut away the deaf ears and wattles, then cut the comb, cutting a certain distance from the back, and then from the front to join this cut, taking especial care not to go too near the skull. Some operators put a finger inside the mouth to get a firm purchase. We should like to see dubbing done away with, leaving these beautiful fowls as nature makes them; but since amateurs and shows will not agree to this, it is best to give directions for dubbing, as an operation bunglingly performed is sure to give unnecessary pain." To save the bird from excessive loss of blood his wattles are usually cut off a week later. Every superfluous piece of flesh and skin should be removed.

The "Henwife" well says: "Why these poor birds are condemned to submit to this cruel operation is a mystery, unfathomable, I suspect, even by the judges themselves. Cock-fighting being forbidden by law, the cocks should, on principle, be left undubbed, as a protest against this brutal amusement. The comb of the Game male bird is as beautifully formed as that of the Dorking; why then rob it of this great ornament? It is asserted that it is necessary to remove the comb to prevent the cocks injuring each other fatally in fighting; but this is not true; a Dorking will fight for the championship as ardently as any Game bird, and yet his comb is spared. Cockerels will not quarrel if kept apart from hens until the breeding season, when they should be separated, and put on their several walks. If pugnaciously inclined I do not believe that the absence of the comb will save the weaker opponent from destruction; therefore I raise my voice for pity, in favour of the beautiful Game cock."

The colours are various, and they are classed into numerous varieties and sub-varieties, of which the chief are—Black-breasted Red; Brown-Red; Silver Duck-wing Greys, so called from the feathers resembling those of a duck; Greys; Blues; Duns; Piles, or Pieds; Black; White; and Brassy-winged, which is Black with yellow on the lesser wing coverts. Colours and markings must be allowed a somewhat wide range in this breed; and figure, with courage, may be held to prove purity of blood though the colour be doubtful. Mr. Douglas considers the Black-breasted Red the finest feathered Game, and states that he never found any come so true to colour as a brood of that variety. White in the tail feathers is highly objectionable, though not an absolute disqualification. White fowls should be entirely white, with white legs. The rules for the coloured legs are very undecided. Light legs match light-coloured birds best. No particular colour is imperative, but it should harmonise with the plumage, and all in a pen must agree.

The best layers are the Black-breasted Reds with willow legs, and the worst the Greys.

CHAPTER XIV.

DORKINGS.

THIS is one of the finest breeds, and especially English. A pure Dorking is distinguished by an additional or fifth toe. There are several varieties, which are all comprised in two distinct classes—the White and the Coloured. The rose-combed white breed is *the* Dorking of the old fanciers, and most probably the original breed, from which the coloured varieties were produced by crossing it with the old Sussex, or some other large coloured fowl. "That such was the case," says Mr. Wright, "is almost proved by the fact that only a few years ago nothing was more uncertain than the appearance of the fifth toe in coloured chickens, even of the best strains. Such uncertainty in any important point is always an indication of mixed blood; and that it was so in this case is shown by the result of long and careful breeding, which has now rendered the fifth toe permanent, and finally established the variety." Mr. Brent says: "The *old* Dorking, the *pure* Dorking, the *only* Dorking, is the *White* Dorking. It is of good size, compact and plump form, with short neck, short white legs, five toes, a full rose-comb, a large breast, and a plumage of spotless white. The practice of crossing with a Game cock was much in vogue with the old breeders, to improve a worn-out stock (which, however, would have been better accomplished by procuring a fresh bird of the same kind, but not related). This cross shows itself in single combs, loss of a claw, or an occasional red feather, but what is still more objectionable, in pale-yellow legs and a yellow circle about the beak, which also indicates a yellowish skin. These, then, are faults to be avoided. As regards size, the White Dorking is generally inferior to the Sussex fowl (or 'coloured Dorking'), but in this respect it only requires attention and careful breeding. The pure White Dorking may truly be considered as fancy stock, as well as useful, because they will breed true to their points; but the grey Sussex, Surrey or Coloured Dorking, often sport. To the breeders and admirers of the so-called 'Coloured Dorkings' I would say, continue to improve the fowl of your choice, but let him be known by his right title; do not support him on another's fame, nor yet deny that the rose-comb or fifth toe is essential to a Dorking, because your favourites are not constant to those points. The absence of the fifth claw to the Dorking would be a great defect, but to the Sussex fowl (erroneously called a 'Coloured Dorking') it is my opinion it would be an improvement, provided the leg did not get longer with the loss."

The fifth toe should not be excessively large, or too far above the ordinary toe.

The White Dorking must have the plumage uniformly white, though in the older birds the hackle and saddle may attain a light golden tint. The rose-comb is preferable, and the beak and legs should be light and clear.

The Coloured Dorking is now bred to great size and beauty. It is a large, plump, compact, square-made bird, with short white legs, and should have a well-developed fifth toe. The plumage is very varied, and may have a wide range, and might almost be termed immaterial, provided a coarse mealy appearance be avoided, and the pen is well matched. This latitude in respect of plumage is so generally admitted that the assertion "you cannot breed Dorkings true to colour," has almost acquired the authority of a proverb. They may be shown with either rose or single combs, but all the birds in a pen must match.

The Dorking is the perfection of a table bird, combining delicately-flavoured white flesh, which is produced in greatest quantity in the choicest parts—the breast, merrythought, and wings—equal distribution of fat, and symmetrical shape. Mr. Baily prefers the Speckled or Grey to the White, as "they are larger, hardier, and fatten more readily; and although it may appear anomalous, it is not less true that white-feathered poultry has a tendency to yellowness in the flesh and fat." Size is an important point in Dorkings. Coloured prize birds weigh from seven to fourteen pounds, and eight months' chickens six or seven pounds. The White Dorking is smaller.

They are not good layers, except when very young, and are bad winter layers. The eggs are large, averaging $2\frac{3}{4}$ ounces, pure white, very much rounded, and nearly equal in size at each end. The hen is an excellent sitter and mother. The chickens are very delicate, requiring more care when young than most breeds, and none show a greater mortality, no more than two-thirds of a brood usually surviving the fourth week of their life. They should not be hatched before March, and must be kept on gravel soil, hard clay, or other equally dry ground, and never on brick, stone, or wooden flooring.

This breed will only thrive on a dry soil. They are fond of a wide range, and cannot be kept within a fence of less than seven feet in height. When allowed unlimited range they appear to grow hardy, and are as easily reared as any other breed if not hatched too early. If kept in confinement they should have fresh turf every day, besides other vegetable food. Dorkings degenerate more than any breed by inter-breeding, and rapidly decrease in size.

Dorkings are peculiarly subject to a chronic inflammation or abscess of the foot, known as "bumblefoot," which probably originated in heavy fowls descending from high perches and walking over sharp stones. The additional toe may have rendered them more liable to this disease. It may now arise from the same cause, and is best prevented by using broad, low perches, and keeping their runs clear of sharp, rough stones, but it also appears to have become hereditary in some birds. There is no cure for it when matured except its removal, and this operation fails oftener than it succeeds; but Mr. Tegetmeier states, that he has in early cases removed the corn-like or wart-like tumours on the ball of the foot with which the disease begins, and cauterised the part with nitrate of silver successfully.

Golden-pencilled and Silver-spangled Hamburghs.   Black Spanish.

CHAPTER XV.

SPANISH.

This splendid breed was originally imported from Spain, and is characterised by its peculiar white face, which in the cock should extend from the comb downwards, including the entire face, and meet beneath in a white cravat, hidden by the wattles; and in the hen should be equally striking. The plumage is perfectly black, with brilliant metallic lustre, reflecting rich green and purple tints. The tail should resemble a sickle in the cock, and be square in the hen. The comb should be of a bright red, large, and high, upright in the cock, but pendent in the hen; the legs blue, clean, and long, and the bearing proud and gallant.

With care they will thrive in a very small space, and are perhaps better adapted for town than any other variety. They are tolerably hardy when grown, but suffer much from cold and wet. Their combs and wattles are liable to be injured by severe cold, from which these fowls should be carefully protected. If frost-bitten, the parts should be rubbed with snow or cold water, and the birds must not be taken into a warm room until recovered.

The Spanish are excellent layers, producing five or six eggs weekly from February to August, and two or three weekly from November to February, and also laying earlier than any other breed except the Brahma, the pullets beginning to lay before they are six months old. Although the hens are only of an average size, and but moderate eaters, their eggs are larger than those of any other breed, averaging $3\frac{1}{2}$ ounces, and some weighing $4\frac{1}{2}$ ounces, each. The shells are very thin and white, and the largest eggs are laid in the spring.

The flesh is excellent, but the body is small compared to

that of the Dorking. They very seldom show any inclination to sit, and if they hatch a brood are bad nurses. The chickens are very delicate, and are best hatched at the end of April and during May. They do not feather till almost three-parts grown, and require a steady mother that will keep with them till they are safely feathered, and therefore the eggs should be set under a Dorking hen, because that breed remains longer with the chicks than any other. They almost always have white feathers in the flight of the wings, but these become black.

"In purchasing Spanish fowls," says an excellent authority, "blue legs, the entire absence of white or coloured feathers in the plumage, and a large white face, with a very large, high comb, which should be erect in the cock, though pendent in the hen, should be insisted on." Legginess is a fault that breeders must be careful to avoid.

The cockerels show the white face earlier than the pullets, and a blue, shrivelly appearance in the face of the chickens is a better sign of future whiteness than a red fleshiness. Pullets are rarely fully whitefaced till above a year old. "The white face," says an excellent authority, "should always extend well around the eye, and up to the point of junction with the comb, though a line of short black feathers is there frequently seen to intrude its undesired presence. It is certainly objectionable, and the less of it the better; but any attempt to remove or disguise this eyesore should be followed by immediate disqualification." Some exhibitors of Spanish shave the down of the edges of the white-face, in order to make it smooth and larger. This disgraceful practice is not allowed at the Birmingham Show.

"One test of condition," says Mr. Baily, "more particularly of the pullets, is the state of the comb, which will be red, soft, and developed, just in proportion to the condition of the bird. While moulting—and they are almost naked during this process—the comb entirely shrivels up."

The White-faced WHITE SPANISH is thought to be merely a sport of the White-faced Black Spanish. But, whatever their origin may have been, they possess every indication of common blood with their Black relatives, and their claims to appear by their side in the exhibition room are as good as those of the White Cochins and the White Polish. The plumage is uniformly white, but in all other respects they resemble the Black breed. From the absence of contrast of colour shown in the face, comb, and plumage of the Black Spanish, the White variety is far less striking in appearance.

The ANDALUSIAN are so called from having been brought from the Spanish province of Andalusia. This breed is of a bluish grey, sometimes slightly laced with a darker shade, but having the neck hackles and tail feathers of a glossy black, with red face and white ears. The chickens are very hardy, and feather well, and earlier than the Spanish.

The MINORCA is so called from having been imported from that island, and is a larger and more compactly-formed breed, resembling the Spanish in its general characteristics; black, with metallic lustre, but with red face, and having only the ear-lobes white; showing even a larger comb, and with shorter legs. They are better as table fowls than the Spanish, but the Andalusian are superior to either. The Minorca is the best layer of all the Spanish breeds, its chickens are tolerably hardy, and it is altogether far superior to the White-faced breed.

ANCONA is a provincial term applied to black and white mottled, or "cuckoo," which on all other points resemble Minorcas, but are smaller.

The "Black Rot," to which Spanish fowls are subject, is a blackening of the comb, swelling of the legs and feet, and general wasting of the system; and can only be cured in the earlier stages by frequent purgings with castor oil, combined with warm nourishing food, and strong ale, or other stimulants, given freely. They are also subject to a peculiar kind of swelled face, which first appears like a small knob under the skin, and increases till it has covered one side of the face. It is considered to be incurable.

CHAPTER XVI.

HAMBURGS.

This breed is medium-sized, and should have a brilliant red, finely-serrated rose-comb, terminating in a spike at the back, taper blue legs, ample tail, exact markings, a well-developed white deaf-ear, and a quick, spirited bearing. They are classed in three varieties, the Pencilled, Spangled, and Black varieties, with the sub-varieties of Gold and Silver in the two former.

The Pencilled Hamburg is of two ground colours, gold and silver, that is, of a brown yellow or white, and very minutely marked. The hens of both colours should have the body clearly pencilled across with several bars of black. The hackle in both sexes should be free from dark marks. In the Golden-pencilled variety the cock should be of one uniform red all over his body without any pencilling whatever, and his tail copper colour; but many first-class birds have pure black tails and the sickle feathers should be shaded with a rich bronze or copper. In the Silver-pencilled variety the cock is often nearly white, with yellowish wing-coverts, and a brown or chestnut patch on the flight feathers of his wing. The tail should be black and the sickle feathers tinged with a reddish white.

The Speckled or Spangled Hamburg, also called Pheasant Fowl, from the false idea that the pheasant was one of its parents, is of two kinds, the Golden-speckled and Silver-speckled, according to their ground colour, the marking taking the form of a spot upon each feather. They have very full double and firmly fixed combs, the point at the end turning upwards, a dark rim round the eyes, blue legs, and mixed hackle. They were also called Moss Fowls, and Mooneys, the latter probably because the end of every feather should have a black rim on the yellow or white ground. In the Golden-spangled some judges prefer cocks with a pure black breast, but others desire them spangled.

"One chief cause of discussion," says Miss Watts, "relating to the Hamburg, regarded the markings on the cocks. The Yorkshire breed, which had been a favourite in that county for many years, produced henny cocks—*i.e.* cocks with plumage resembling that of a hen. The feathers of the hackle were not narrow and elongated like those of cocks generally, but were short and rounded like those of the hen; the saddle-feathers were the same, and the tail, instead of being graced with fine flowing sickle feathers, was merely square like that of a hen. The Lancashire Mooneys, on the contrary, produce cocks with as fine flowing plumage as need grace any chanticleer in the land, and tails with sickle-feathers twenty-two inches long, fine flowing saddle-feathers, and abundant hackle. The hen-tail cocks had the markings, as well as the form, of the hen; the long feathers of the others cannot, from their form, have these markings. On this question party-spirit ran high: York and Lancaster, Cavalier and Roundhead, were small discussions compared with it; but the hen-cocks were beaten, and we now seldom hear of them. A mixture of the two breeds has been tried; but by it valuable qualities and purity of race have been sacrificed."

The Black Hamburg is of a beautiful black with a metallic lustre, and is a noble-looking bird, the cocks often weighing seven pounds. There is little doubt that it was produced by crossing with the Spanish, which blood shows itself in the white face, which is often half apparent, and in the darker legs. But it is well established as a distinct variety, and good birds breed true to colour and points. The cocks' combs are larger, and the hens' legs shorter, than the other varieties.

Bolton Bays and Greys, Chitteprats, Turkish, and Creoles or Corals, Pencilled Dutch fowls, and Dutch every-day layers, are but incorrect names for the Hamburgs, with which they are identical.

The Hamburgs do not attain to their full beauty until

three years old. "As a general rule," says Mr. Baily, "no true bred Hamburg fowl has top-knot, single comb, white legs, any approach to feather on the legs, white tail, or spotted hackle." The white ear-lobe being so characteristic a feature in all the Hamburgs, becomes most important in judging their merits. Weight is not considered, but still the Pencilled cock should not weigh less than four and a half pounds, nor the hen than three and a half; and the Spangled cock five pounds and the hen four.

The Hamburgs are most prolific layers naturally, without over-stimulating feeding, surpassing all others in the number of their eggs, and deserve their popular name of "everlasting layers." Their eggs are white, and do not weigh more than $1\frac{1}{2}$ ounce to $1\frac{3}{4}$ ounce each; and the hens are known to average 240 eggs yearly. Not being large eaters, they are very profitable fowls to keep. The eggs of the Golden-spangled are the largest, and it is the hardiest variety, but the Pencilled lay more. The Black variety produces large eggs, and lays a greater number than any known breed.

They very seldom show any desire to sit except when they have a free woodland range, for even if free it must be wild to induce any desire to perpetuate the species, and they never sit if confined to a yard. The chickens should not be hatched earlier than May, but in the South of England they will do very well if hatched by a Cochin-China hen at the beginning of March. They are small birds for table, but of excellent quality.

Hamburgs do not bear confinement well, and will not thrive without a good run; a grass field is the best. Being small and light, even a ten-feet fence will not keep them within a small run. They may indeed be kept in a shed, but the number must be very few in proportion to its size, and they must be kept dry and scrupulously clean. They are excellent guards in the country, for if disturbed in their roosting-place they will make a great noise. The breed has improved in this country, and British bred fowls are much stronger than the imported birds.

Golden and Silver-spangled.

Polish.

White-crested Black.

CHAPTER XVII.

POLANDS.

This breed might with good reason be divided into more families, but it is usual to rank as Polands all fowls with their chief distinguishing characteristic, a full, large, round, compact tuft on the head. The breed "is quite unknown in Poland, and takes its name," says Mr. Dickson, "from some resemblance having been fancied between its tufted crest and the square-spreading crown of the feathered caps worn by the Polish soldiers." It is much esteemed in Egypt, and equally abundant at the Cape of Good Hope, where their legs are feathered. Some travellers assert that the Mexican poultry are crested, and that what are called Poland fowls are natives of either Mexico or South America; but others believe that they are natives of the East, and that they, as well as all the other fowls on the Continent of America, have been introduced from the Old World.

The Golden-spangled and Silver-spangled are the most beautiful varieties, the first being of a gold colour and the second white, both spangled with black. The more uniform the colour of the tuft is with that of the bird, the higher it is valued.

The Black Poland is of a deep velvety black; has a large, white, round tuft, and should not have a comb, but many have a little comb in the form of two small points before the tuft. The tuft to be perfect should be entirely white, but it is rare to meet with one without a slight bordering of black, or partly black, feathers round the front.

There are also Yellow, laced with white, Buff or Chamois, spangled with white, Blue, Grey, Black, and White mottled. All the sub-varieties should be of medium size, neat compact form, plump, full-breasted, and have lead-coloured legs and; ample tails.

The top-knot of the cock should be composed of straight feathers, growing from the centre of the crown, and falling over outside, but not so much as to intercept the sight, and form a circular crest. That of the hen should be formed of feathers growing out and turning in at the extremity, so as to resemble a cauliflower, and it should be even, firm, and as nearly round as possible. Large, uneven top-knots composed of loose feathers do not equal smaller but firm and well-shaped crests. The white ear-lobe is essential in all the varieties.

"Beards" in Polands were formerly not admired. Among the early birds brought from the continent, not one in a hundred was bearded, and those that were so were often rejected, and it was a question of dispute whether the pure bird should have them or not. Bearded birds at shows were the exceptions, but an unbearded pen of Polands is now seldom or ever seen.

There was formerly a breed of White, with black top-knots, but that is lost, although it seems to have been not only the most ornamental, but the largest and most valuable of all the Polish varieties. The last specimen known was seen by Mr. Brent at St. Omer in 1854, and it is possible that the breed may still exist in France or Ireland.

The SERAI TA-OOK, or FOWL OF THE SULTAN, is the latest Polish fowl introduced into this country. They were imported in 1854 by Miss Watts, who says: "With regard to the name, Serai is the name of the Sultan's palace; Tä-ook is Turkish for fowl; the simplest translation of this is, Sultan's fowls, or fowls of the Sultan; a name which has the double advantage of being the nearest to be found to that by which they have been known in their own country, and of designating the country from which they came. In general habits they are brisk and happy-tempered, but not kept in as easily as Cochin-Chinas. They are very good layers; their eggs are large and white; they are non-sitters, and small eaters. A grass run with them will remain green long after the crop would have been cleared by either

Brahmas or Cochins, and with scattered food they soon become satisfied and walk away. They are the size of our English Poland fowls. Their plumage is white and flowing; they have a full-sized compact Poland tuft on the head, are muffed, have a good flowing tail, short well-feathered legs, and five toes upon each foot. The comb is merely two little points, and the wattles very small. We have never seen fowls more fully decorated—full tail, abundant furnishing, in hackle almost touching the ground, boots, vulture-hocks, beards, whiskers, and full round Poland crests. Their colour is pure white."

They are prolific layers during spring and summer. Their eggs are white, and weigh from 2 ounces to $2\frac{1}{4}$ ounces each, the Spangled varieties producing the largest. They rarely sit, and generally leave their eggs after five or six days, and are not good mothers. The chickens require great care for six weeks. They should never be hatched by heavy hens, as the prominence in the skull which supports the top-knot is never completely covered with bone, and very sensible to injury. Like the Game breed they improve in feather for several years. Polands never thrive on a wet or cold soil, and are more affected by bad weather than any other breed; the top-knots being very liable to be saturated with wet. They are easily fattened, and their flesh is white, juicy, and rich-flavoured, but they are not sufficiently large for the market.

Mr. Hewitt cautions breeders against attempting to seize birds suddenly, as the crest obscures their sight, and, being taken by surprise, they are frequently so frightened as to die in the hand. They should, therefore, always be spoken to, or their attention otherwise attracted before being touched.

CHAPTER XVIII.

BANTAMS.

OF this breed one kind is Game, and resembles the Game fowl, except in size; another is feathered to the very toes, the feathers on the tarsi, or beam of the leg, being long and stiff, and often brushing the ground. They are peculiarly fancy fowls. There are several varieties, the White, Black, Nankin, Partridge, Booted or Feather-legged, Game, and the Golden-laced and Silver-laced, or Sebright Bantam. All should be very small, varying from fourteen to twenty ounces in the hen, and from sixteen to twenty-four in the cock. The head should be narrow; beak curved; forehead rounded; eyes bright; back short; body round and full; breast very prominent; legs short and clean, except in the Booted variety; wings depressed; and the carriage unusually erect, the back of the neck and the tail feathers almost touching; and the whole bearing graceful, bold, and proud.

"The Javanese jungle-fowl" (*Gallus Bankiva*), says Mr. W. C. L. Martin, "the Ayam-utan of the Malays, is a native of Java; but either a variety or a distinct species of larger size, yet very similar in colouring, is found in continental India. The Javanese, or Bankiva jungle-fowl, is about the size of an ordinary Bantam, and in plumage resembles the black-breasted red Game-bird of our country, with a steel-blue mark across the wings. The comb is high, its edge is deeply serrated, and the wattles are rather large. The hackle feathers of the neck and rump are long and of a glossy golden orange; the shoulders are chestnut red, the greater wing-coverts deep steel-blue, the quill feathers brownish black, edged with pale, reddish yellow, or sandy red. The tail is of a black colour, with metallic reflections of green and blue. The under parts are black

Student's Cud and Sweethe...
BANTAMS.

the naked space round the eyes, the comb, and wattles are scarlet. The hen closely resembles a brown hen of the Game breed, except in being very much smaller. That this bird, or its continental ally, is one of the sources—perhaps the main source—of our domestic race, cannot be doubted. It inter-breeds freely with our common poultry, and the progeny is fertile. Most beautiful cross-breeds between the Bankiva jungle-fowl and Bantam may be seen in the gardens of the Zoological Society."

"That the Bankiva jungle-fowl of Java, or its larger continental variety, if it be not a distinct species (and of which Sir W. Jardine states that he has seen several specimens), is one of the sources of our domestic breeds, cannot, we think, be for a moment doubted. It would be difficult to discover any difference between a clean-limbed, black-breasted red Bantam-cock, and a cock Bankiva jungle-fowl. Indeed, the very term Bantam goes far to prove their specific identity. Bantam is a town or city at the bottom of a bay on the northern coast of Java; it was first visited by the Portuguese in 1511, at which time a great trade was carried on by the town with Arabia, Hindostan, and China, chiefly in pepper. Subsequently it fell into the hands of the Dutch, and was at one time the great rendezvous for European shipping. It is now a place of comparative insignificance. From this it would seem that the jungle-fowls domesticated and sold to the Europeans at Bantam continued to be designated by the name of the place where they were obtained, and in process of time the name was appropriated to all our dwarfish breeds."

Game Bantams are exact miniatures of real Game fowls, in Black-breasted red, Duck-wing, and other varieties. The cocks must not have the strut of the Bantam, but the bold, martial bearing of the Game cock. Their wings should be carried closely, and their feathers be hard and close. The Duck-wing cock's lower wing-coverts should be marked with blue, forming a bar across each wing.

The SEBRIGHT, or GOLD AND SILVER-LACED BANTAM, is a breed with clean legs, and of most elegantly spangled plumage, which was bred and has been brought to great perfection by Sir John Sebright, after whom they are named. The attitude of the cock is singularly bold and proud, the head being often thrown so much back as to meet the tail feathers, which are simple like those of a hen, the ordinary sickle-like feathers being abbreviated and broad. The Gold-laced Sebright Bantams should have golden brownish-yellow plumage, each feather being bordered with a lacing of black; the tail square like that of the hen, without sickle feathers, and carried well over the back, each feather being tipped with black, a rose-comb pointed at the back, the wings drooping to the ground, neither saddle nor neck hackles, clean lead-coloured legs and feet, and white ear-lobes; and the hen should correspond exactly with him, but be much smaller. The Silver-laced birds have exactly the same points except in the ground feathering, which should be silvery, and the nearer the shade approaches to white the more beautiful will be the bird. Their carriage should resemble that of a good Fantail pigeon.

The BLACK BANTAMS should be uniform in colour, with well-developed white ear-lobes, rose-combs, full hackles, sickled and flowing tail, and deep slate-coloured legs. The WHITE BANTAMS should have white legs and beak. Both should be of tiny size.

The NANKIN, or COMMON YELLOW BANTAM, is probably the nearest approach to the original type of the family—the "Bankiva fowl." The cock "has a large proportion of red and dark chestnut on the body, with a full black tail; while the hen is a pale orange yellow, with a tail tipped with black, and the hackle lightly pencilled with the same colour, and clean legs. Combs vary, but the rose is decidedly preferable. True-bred specimens of these birds being by no means common, considerable deviations from the above description may consequently be expected in birds passing under this appellation."

The BOOTED BANTAMS have their legs plumed to the toes, not on one side only like Cochin-Chinas, but completely on both, with stiff, long feathers, which brush the ground. The most beautiful specimens are of a pure white.

"Feathered-legged Bantams," says Mr. Baily, "may be of any colour; the old-fashioned birds were very small, falcon-hocked, and feathered, with long quill feathers to the extremity of the toe. Many of them were bearded. They are now very scarce; indeed, till exhibitions brought them again into notice, these beautiful specimens of their tribe were all neglected and fast passing away. Nothing but the Sebright was cultivated; but now we bid fair to revive the pets of our ancestors in all their beauty."

The PEKIN, or COCHIN BANTAMS, were taken from the Summer Palace at Pekin during the Chinese war, and brought to this country. They exactly resemble the Buff Cochins in all respects except size. They are very tame.

The JAPANESE BANTAM is a recent importation, and differs from most of the other varieties in having a very large single comb. It has very short well-feathered legs, and the colour varies. Some are quite white, some have pure white bodies, with glossy, jet-black tails, others are mottled and buff. They throw the tail up and the head back till they nearly meet, as in the Fantailed pigeon. They are said to be the constant companions of man in their native country, and have a droll and good-natured expression.

All the Bantam cocks are very pugnacious, and though the hens are good mothers to their own chickens, they will attack any stranger with fury. They are good layers of small but exquisitely-flavoured eggs. But no breed produces so great a proportion of unfertile eggs. June is the best month for hatching, as the chickens are delicate. They feather more quickly than most breeds, and are apt to die at that period through the great drain upon the system in producing feathers. When fully feathered they are quite hardy. The hens are excellent mothers. The chickens require a little more animal food than other fowls, and extra attention for a week or two in keeping them dry. Bantams are very useful in a garden, eating many slugs and insects, and doing little damage.

CHAPTER XIX.

FRENCH AND VARIOUS.

The French breeds are remarkable for great weight and excellent quality of flesh, with a very small proportion of bones and offal; their breeders having paid great attention to those important, substantial, and commercial points instead of devoting almost exclusive attention to colour and other fancy points as we have done. As a rule they are all non-sitters, or sit but rarely.

The Crève Cœur has been known the longest and most generally. This breed is said to derive its name from a village so called in Normandy, whence its origin can be distinctly traced; but others fancifully say, from the resemblance of its peculiar comb to a broken heart. It is scarce, and pure-bred birds are difficult to procure. The Crève Cœur is a fine large bird, black in plumage, or nearly so, with short, clean black legs, square body, deep chest, and a large and extraordinary crest or comb, which is thus described by M. Jacque: "Various, but always forming two horns, sometimes parallel, straight, and fleshy; sometimes joined at the base, slightly notched, pointed, and separating at their extremities; sometimes adding to this latter description interior ramifications like the horns of a young stag. The comb, shaped like horns, gives the Crève Cœur the appearance of a devil." It is bearded, and has a top-knot or crest behind the comb. They are very quiet, walk slowly, scratch but little, do not fly, are very tame, ramble but little, and prefer seeking their food on the dunghill in the poultry-yard to wandering afar off. They are the most contented of all breeds in confinement, and will thrive in a limited space. They are tame, tractable fowls, but inclined to roup and similar diseases in our climate, and therefore prosper most on a dry, light soil,

and can scarcely have too much sun. They are excellent layers of very large white eggs.

The chickens grow so fast, and are so inclined to fatten, that they may be put up at from ten to twelve weeks of age, and well fattened in fifteen days. The Crêve Cœur is a splendid table bird, both for the quantity and quality of its flesh. The hen is heavy in proportion to the cock, weighing eight and a half pounds against his nine and a half, and the pullets always outweigh the cockerels.

LA FLECHE is thus described by M. Jacque: "A strong, firm body, well placed on its legs, and long muscular feet, appearing less than it really is, because the feathers are close; every muscular part well developed; black plumage. The La Flêche is the tallest of all French cocks; it has many points of resemblance with the Spanish, from which I believe it to be descended by crossing with the Crêve Cœur. Others believe that it is connected with the Brêda, which it does, in fact, resemble, in some particulars. It has white, loose, and transparent skin; short, juicy, and delicate flesh, which puts on fat easily."

"The comb is transversal, double, forming two horns bending forward, united at their base, divided at their summits, sometimes even and pointed, sometimes having ramifications on the inner sides. A little double 'combling' protrudes from the upper part of the nostrils, and although hardly as large as a pea, this combling, which surmounts the sort of rising formed by the protrusion of the nostrils, contributes to the singular aspect of the head. This measured prominence of the comb seems to add to the characteristic depression of the beak, and gives the bird a likeness to a rhinoceros." The plumage is jet black, with a very rich metallic lustre; large ear-lobe of pure white; bright red face, unusually free from feathers; and bright lead-coloured legs, with hard, firm scales. They are very handsome, showy, large, and lively birds, more inclined to wander than the Crêve Cœur, and hardier when full grown; but their chickens are even more delicate in wet weather, and should not be hatched before May. They are easily reared, and grow quickly. They are excellent layers of very large white eggs, but do not lay well in winter, unless under very favourable circumstances, and resemble the Spanish in the size and number of their eggs, and the time and duration of laying. Their flesh is excellent, juicy, and resembles that of the Game fowl, and the skin white and transparent, but the legs are dark. This breed is larger and has more style than the Crêve Cœur, and is better adapted to our climate; but the fowls lack constitution, particularly the cocks, and are very liable to leg weakness and disease of the knee-joint, and when they get out of condition seldom recover. They are found in the north of France, but are not common even there.

The HOUDAN has the size, deep compact body, short legs, and fifth toe of the Dorking. They are generally white, some having black spots as large as a shilling, are bearded, and should have good top-knots of black and white feathers, falling backwards like a lark's crest; and the remarkable comb is thus described by M. Jacque: "Triple, transversal in the direction of the beak, composed of two flattened spikes, of long and rectangular form, opening from right to left, like two leaves of a book; thick, fleshy, and variegated at the edges. A third spike grows between these two, having somewhat the shape of an irregular strawberry, and the size of a long nut. Another, quite detached from the others, about the size of a pea, should show between the nostrils, above the beak."

Mr. F. H. Schröder, of the National Poultry Company, considered that this surpassed all the French breeds, combining the size, shape, and quality of flesh of the Dorking with earlier maturity; prolific laying of good-sized eggs, which are nearly always fertile, and on this point the opposite of the Dorking; and early and rapid feathering in the chickens, which are, notwithstanding, hardier than any breeds except the Cochin and Brahma. They are very hardy, never sick, and will thrive in a small space. They are smaller than the Crêve Cœur or La Flêche, but well shaped and plump; and for combining size and quality of flesh with quantity and size of eggs nothing can surpass them.

SCOTCH DUMPIES, GO LAIGHS, BAKIES, or CREEPERS, are almost extinct; but they are profitable fowls, and ought to be more common, as they are very hardy, productive layers of fine large eggs, and their flesh is white and of excellent quality. They should have large, heavy bodies; short, white, clean legs, not above an inch and a half or two inches in length. The plumage is a mixture of black or brown, and white. They are good layers of fine large eggs. They cannot be surpassed as sitters and mothers, and are much valued by gamekeepers for hatching the eggs of pheasants. The cocks should weigh six or seven and the hen five or six pounds.

The SILKY fowl is so called from its plumage, which is snowy white, being all discomposed and loose, and of a silky appearance, resembling spun glass. The comb and wattles are purple; the bones and the periosteum, or membrane covering the bones, black, and the skin blue or purple; but the flesh, however, is white and tender, and superior to that of most breeds. It is a good layer of small, round, and excellent eggs. The cock generally weighs less than three, and the hen less than two, pounds. It comes from Japan and China, and generally thrives in our climate. The chickens are easily reared if not hatched before April nor later than June. They are capital foster mothers for partridges, and other small and tender game.

The RUMPKIN, or RUMPLESS fowl, a Persian breed, not only lacks the tail-feathers but the tail itself. It is hardy, of moderate size, and varies in colour, but is generally black or brown, and from the absence of tail appears rounder than other fowls. The hens are good layers, but the eggs are often unfertile. They are good sitters and mothers, and the flesh is of fair quality.

The FRIESLAND, so named from confounding the term "frizzled" with Friesland, is remarkable from having all the feathers, except those of the wings and tail, frizzled, or curled up the wrong way. It is small, very delicate, and a shower drenches it to the skin.

BARNDOOR fowl are a mongrel race, compounded by chance, usually of the Game, Dorking, and Polish breeds.

CHAPTER XX.

TURKEYS.

TURKEYS are not considered profitable except on light, dry soils, which is said to be the cause of their success in Norfolk. They prosper, however, in Ireland ; but although the air there is moist, the soil is dry, except in the boggy districts. Miss Watts believes that "any place in which turkeys are properly reared and fed may compete with Norfolk. Very fine birds may be seen in Surrey, and other places near London." The general opinion of the best judges is, that they can barely be made to repay the cost of their food, which is doubtless owing to the usual great mortality among the chicks, which loss outbalances all profit ; but others make them yield a fair profit, simply because, from good situation and judicious management, they rear all, or nearly all, the chicks. A single brood may be reared with ease on a small farm or private establishment without much extra expense, where sufficient attention can be devoted to them ; but to make them profitable they should be bred on a large scale, and receive exclusive attention. They should have a large shed or house, with a boarded floor, to themselves.

Turkeys must have space, for they are birds of rambling habits, and only fitted for the farmyard, or extensive runs, delighting to wander in the fields in quest of insects, on which, with green herbage, berries, beech-mast, and various seeds, they greedily feed. The troop will ramble about all day, returning to roost in the evening, when they should have a good supply of grain ; and another should be given in the morning, which will not only induce them to return home regularly every night, but keep them in good store. condition, so that they can at any time be speedily fattened. Peas, vetches, tares, and most sorts of pulse, are almost

poisonous to them. Their feeding-place must be separate from the other poultry, or they will gobble up more than their share. Turkeys will rarely roost in a fowl-house, and should have a very high open shed, the perches being placed as high as possible. They are extremely hardy, roosting, if allowed, on the highest trees in the severest weather. But this should be prevented, as their feet are apt to become frost-bitten in severe weather. The chickens are as delicate: Wet is fatal to them, and the very slightest shower even in warm weather will frequently destroy half a brood.

The breeding birds should be carefully selected, any malformation almost invariably proving itself hereditary. The cock is at maturity when a year old, but not in his prime till he has attained his third year, and is entering upon his fourth, and he continues in vigour for three or four years more. He should be vigorous, broad-breasted, clean-legged, with ample wings, well-developed tail, bright eyes, and the carunculated skin of the neck full and rapid in its changes of colour. The largest possible hen should be chosen, the size of the brood depending far more upon the female than the male. One visit to the male is sufficient to render all the eggs fertile, and the number of hens may be unlimited, but to obtain fine birds, twelve or fifteen hens to one cock is the best proportion. The hen breeds in the spring following that in which she was hatched, but is not in her prime till two or three years old, and continues for two or three years in full vigour.

The hen generally commences laying about the middle of March, but sometimes earlier. When from her uttering a peculiar cry and prying about in quest of a secret spot for sitting, it is evident that she is ready to lay, she should be confined in the shed, barn, or other place where the nest has been prepared for her, and let out when she has laid an egg. The nest should be made of straw and dried leaves, in a large wicker basket, in a quiet secluded place, and an egg or nest-egg of chalk should be placed in it to induce her to adopt it. Turkeys like to choose their own laying-places, and keep to them though their eggs are removed daily, provided a nest-egg is left there. They will wander to a distance in search of a secluded spot for laying, and pay their visits to the nest so cleverly that sometimes they keep it a secret and hatch a brood there, which, however, does not generally prove a strong or large one as in the case of ordinary fowls. When a hen has chosen a safe, quiet, and sheltered place for her nest, it is best to give her more eggs when she shows a desire to sit, and let her stay there. The hen generally lays from fifteen to twenty eggs, sometimes fewer and often many more. As soon as seven are produced, they should be placed under a good common hen, a Cochin is the best, and the remainder can be put under her when she wants to sit. The best hatching period is from the end of March to May, and none should be hatched later than June. The broody hens may be placed on their eggs in any quiet place, as they are patient, constant sitters, and will not leave their eggs wherever they may be put. A hen may be allowed from nine to fifteen eggs, according to her size. During the time the hen is sitting she requires constant attention. She must occasionally be taken off the nest to feed, and regularly supplied with fresh water; otherwise she will continue to sit without leaving for food, till completely exhausted. In general, do not let the cock go near the sitting hen, or he will destroy the eggs or chicks; but some behave well, and may be left at large with safety. She should not be disturbed or visited by any one but the person she is accustomed to be fed by, and the eggs should not be touched unnecessarily.

The chickens break the shell from the twenty-sixth to the twenty-ninth day, but sometimes as late as the thirty-first. Let them remain in the nest for twenty-four hours, but remove the shells, and next morning place the hen under a roomy coop or crate, on boards, in a warm outhouse. Keep her and her brood cooped up for two months, moving the coop every fine day into a dry grass field, but keep them in an outhouse in cold or wet weather. The chicks having a great tendency to diarrhœa, the very best food for the first week is hard-boiled eggs, chopped small, mixed with minced dandelion, and when that cannot be had, with boiled nettles. They may then have boiled egg, bread-crumbs, and barleymeal for a fortnight, when the egg may be replaced by boiled potato, and small grain may soon be added. Do not force them to eat, but give them a little food on the tip of your finger, and they will soon learn to pick it out of the trough. A little hempseed, suet, onion-tops, green mustard, and nettle-tops, chopped very fine, should be mixed with their food. Curds are excellent food, and easily prepared by mixing powdered alum with milk slightly warmed, in the proportion of one teaspoonful of alum to four quarts of milk, and, when curdled, separating the curds from the whey. They should be squeezed very dry, and must always be given in a soft state. Water should be given but sparingly, and never allowed to stand by them, but when they have had sufficient it should be taken or thrown away. The water must be put in pans so contrived or placed that they cannot wet themselves. (*See* page 38.) Fresh milk is apt to disagree with the young chicks, and is not necessary. If a chick shows weakness, or has taken cold, give it some carraway seeds.

In their wild state the turkey rears only one brood in a season, and it is not advisable to induce the domesticated bird by any expedients to hatch a second, for it would be not only detrimental to her, but the brood would be hatched late in the season, and very difficult to rear, while those reared would not be strong, healthy birds.

The coop should be like that used for common fowls, but two feet broad, and higher, being about three feet high in front and one foot at the back; this greater slant of the roof being made in order to confine her movements, as otherwise she would move about too much, and trample upon her brood. When they have grown larger they must have a larger coop, made of open bars wide enough apart for them to go in and out, but too close to let in fowls to eat their delicate food, and the hen must be placed under it with them. A large empty crate, such as is used to contain crockery-ware, will make a good coop for large poults; but if one cannot be had, a coop may be made of laths or rails, with the bars four inches apart; it should be about five feet long, four feet broad, and three feet high.

Keep her cooped for two months, moving the coop every fine, dry day into a grass field, but on cold or wet days keep them in the outhouse. If she is allowed her liberty before they are well grown and strong, she will wander away with them through the long grass, hedges, and ditches, over highway, common, and meadow, mile after mile, losing them on the road, and straying on with the greatest complacency, and perfectly satisfied so long as she has one or two following her, and never once turning her head to see how her panting chicks are getting on, nor troubled when they squat down tired out, and implore her plaintively to come back; and all this arises from sheer heedlessness, and not from want of affection, for she will fight for her brood as valiantly as any pheasant will for hers. When full grown they should never be allowed to roam with her while there is heavy dew or white frost on the grass, but be kept in till the fields and hedgerows are dry. They will pick up many seeds and insects while wandering about in the fields with her, but must be fed by hand three or four times a day at regular intervals.

They cease to be chicks or chickens, and are called turkey-poults when the male and female distinctive characteristics are fairly established, the carunculated skin and comb of the cock being developed, which is called "shooting the red," or "putting out the red," and begins when they are eight or ten weeks old. It is the most critical period of their lives—much more so than moulting, and during the process their food must be increased in quantity, and made more nourishing by the addition of boiled egg-yolks, bread crumbled in ale, wheaten flour, bruised hempseed, and the like, and they must be well housed at night. When this process is completed they will be hardy, and able to take care of themselves; but till they are fully fledged it will be advisable to keep them from rain and cold, and not to try their hardness too suddenly.

Vegetables, as chopped nettles, turnip-tops, cabbage sprouts, onions, docks, and the like, boiled down and well mixed with barleymeal, oatmeal, or wheaten flour, and curds, if they can be afforded, form excellent food for the young poults; also steamed potatoes, boiled carrots, turnips, and the like. With this diet may be given buckwheat, barley, oats, beans, and sunflower seeds.

When they are old enough to be sent to the stubble and fields, they are placed in charge of a boy or girl of from twelve to fifteen years old, who can easily manage one hundred poults. They are driven with a long bean stick, and the duties of the turkey-herd is to keep the cocks from fighting, to lead them to every place where there are acorns, beech-mast, corn, wild fruit, insects, or other food to be picked up. He must not allow them to get fatigued with too long rambles, as they are not fully grown, and must shelter them from the burning sun, and hasten them home on the approach of rain. The best times for these rambles are from eight to ten in the morning, when the dew is off the grass, and from four till seven in the evening, before it begins to fall.

Turkeys are crammed for the London markets. The process of fattening may commence when they are six months old, as they require a longer time to become fit for the market than fowls. The large birds which are seen at Christmas are usually males of the preceding year, and about twenty months old. All experienced breeders repudiate "cramming." To obtain fine birds the chickens must be fed abundantly from their birth until they are sent to market, and while they are being fattened they should be sent to the fields and stubble for a shorter time daily, and their food must be increased in quantity and improved in quality. Early hatched, well fed young Norfolk cocks will frequently weigh twenty-three pounds by Christmas of the same year, and two-year-old birds will sometimes attain to twenty pounds. When two or more years old they are called "stags."

The domesticated turkey can scarcely be said to be divided into distinct breeds like the common fowl, the several varieties being distinguished by colour only, but identical in their form and habits. They vary considerably in colour—some being of a bronzed black, others of a coppery tint, of a delicate fawn colour, or buff, and some of pure white. The dark coloured birds are generally considered the most hardy, aud are usually the largest. The chief varieties are the Cambridge, Norfolk, Irish, American, and French.

The Cambridge combines enormous size, a tendency to fatten speedily, and first-rate flavour. The tortoiseshell character of its plumage gives the adult birds a very prepossessing appearance around the homestead, and a striking character in the exhibition room. The colours may vary from pale to dark grey, with a deep metallic brown tint, and light legs. The legs should be stout and long.

The Norfolk breed is more compact and smaller-boned, and produces a large quantity of meat of delicate whiteness and excellent quality. The cocks are almost as heavy as the Cambridge breed, but the hens are smaller and more compact. The Norfolk should be jet, not blue black, and free from any other colour, being uniform throughout, including the legs and feet.

All the birds in a pen must be uniform.

The American wild turkey has become naturalised in this country, but being of a very wandering disposition is best adapted to be kept in parks and on large tracts of wild land. It is slender in shape, but of good size, with uniform metallic bronze plumage, the flight feathers being barred with white, and the tail alternately with white, rich dark brown, and black, and with bright pink legs. The wattles are smaller than in the other breeds, and of a bluish tinge. They are very hardy, but more spiteful than others, and are said to be also more prolific. Crosses often take place in America between the wild and tame races, and are highly valued both for their appearance and for the table. Eggs of the wild turkey have also often been taken from their nests, and hatched under the domesticated hen. The flavour of the flesh of the American breed is peculiar and exceedingly good, but they do not attain a large size.

CHAPTER XXI.

GUINEA-FOWLS.

The Guinea-fowl, Gallina, or Pintado (*Numida Meleagris*), is the true meleagris of the ancients, a term generically applied by Belon, Aldrovandus, and Gesner to the turkey, and now retained, although the error is acknowledged, in order to prevent confusion. It is a native of Africa, where it is extensively distributed. They associate in large flocks and frequent open glades, the borders of forests, and banks of rivers, which offer abundant supplies of grain, berries, and insects, in quest of which they wander during the day, and collect together at evening, and roost in clusters on the branches of trees or shrubs. Several other wild species are known, some of which are remarkable for their beauty; but the common Guinea-fowl is the only one domesticated in Europe. The Guinea-fowl is about twenty-two inches long, and from standing high on its legs, and having loose, full plumage, appears to be larger than it really is, for when plucked it does not weigh more than an ordinary Dorking. It is very plump and well-proportioned. The Guinea-fowl is not bred so much as the turkey in England or France, is very rare in the northern parts of Europe, and in India is bred almost exclusively by Europeans, although it thrives as well there as in its native country. It "is turbulent and restless," says Mr. Dickson, "continually moving from place to place, and domineering over the whole poultry-yard, boldly attacking even the fiercest turkey cock, and keeping all in alarm by its petulant pugnacity"; and the males, although without spurs, can inflict serious injury on other poultry with their short, hard beaks. The Guinea-fowls make very little use of their wings, and if forced to take to flight, fly but a short distance, then alight, and trust to their rapid mode of running, and their dexterity in threading the mazes of brushwood and dense herbage, for security. They are shy, wary, and alert.

It is not much kept, its habits being wandering, and requiring an extensive range, but as it picks up nearly all its food, and is very prolific, it may be made very profitable in certain localities. The whole management of both the young and the old may be precisely the same as that of turkeys, in hatching, feeding, and fattening. This "species," says Mr. Dickson, "differs from all other poultry, in its being difficult to distinguish the cock from the hen, the chief difference being in the colour of the wattles, which are more of a red hue in the cock, and more tinged with blue in the hen. The cock has also a more stately strut."

They mate in pairs, and therefore an equal number of cocks and hens must be kept, or the eggs will prove unfertile. To obtain stock, some of their eggs must be procured, and placed under a common hen; for if old birds are bought, they will wander away for miles in search of their old home, and never return. They should be fed regularly, and must always have one meal at night, or they will scarcely ever roost at home. They will not sleep in the fowl-house, but prefer roosting in the lower branches of a tree, or on a thick bush, and retire early. They make a peculiar, harsh, querulous noise, which is oft-repeated, and not agreeable. The hens are prolific layers, beginning in May, and continuing during the whole summer. Their eggs are small, but of excellent flavour, of a pale yellowish red, finely dotted with a darker tint, and remarkable for the hardness of the shell. The hen usually lays on a dry bank, in secret places; and a hedgerow a quarter of a mile off is quite as likely to contain her nest as any situation nearer her home. She is very shy, and, if the eggs are taken from her nest, will desert it, and find another; a few should, therefore, always be left, and it should never be visited when she is in sight. But she often contrives to elude all watching, and hatch a brood, frequently at a late period, when the weather is too cold for the chickens. As the Guinea-fowl seldom shows much disposition to incubate if kept under restraint, and frequently sits too late in the season to rear a brood in this country, it is a general practice to place her eggs under a common fowl—Game and Bantams are the best for the purpose. About twenty of the earliest eggs should be set in May. The Guinea-hen will hatch another brood when she feels inclined. They sit for twenty-six to twenty-nine or thirty days. When she sits in due season she generally rears a large brood, twenty not being an unusual number.

The chickens are very tender, and should not be hatched too early in spring, as a cold March wind is generally fatal to them. They must be treated like those of the turkey, and as carefully. They should be fed almost immediately, within six hours of being hatched, abundantly, and often; and they require more animal food than other chickens. Egg boiled hard, chopped very fine, and mixed with oatmeal, is the best food. They will die if kept without food for three or four hours; and should have a constant supply near them until they are allowed to have full liberty and forage for themselves. They will soon pick up insects, &c., and will keep themselves in good condition with a little extra food. They are very strong on their legs, and those hatched under common hens may be allowed to range with her at the end of six weeks, and be fed on the same food and at the same times as other chickens.

The Guinea-fowl may be considered as somewhat intermediate between the pheasant and turkey. After the pheasant season, young birds that have been hatched the same year are excellent substitutes for that fine game, and fetch a fair price. They should never be fattened, but have a good supply of grain and meal for a week or two before being killed. The flesh of the young bird is very delicate, juicy, and well-flavoured, but the old birds, even of the second year, are dry, tough, and tasteless.

CHAPTER XXII.

DUCKS.

DUCKS will not pay if all their food has to be bought, except it is purchased wholesale, and they are reared for town markets, for their appetites are voracious, and they do not graze like geese. They may be kept in a limited space, but more profitably and conveniently where they have the run of a paddock, orchard, kitchen garden, flat common, green lane, or farmyard, with ditches and water. They will return at night, and come to the call of the feeder. Nothing comes amiss to them—green vegetables, especially when boiled, all kinds of meal made into porridge, all kinds of grain, bread, oatcake, the refuse and offal of the kitchen, worms, slugs, snails, insects and their larvæ, are devoured eagerly. Where many fowls are kept, a few ducks may be added profitably, for they may be fed very nearly on what the hens refuse.

Ducks require water to swim in, but "it is a mistake," says Mr. Baily, "to imagine that ducks require a great deal of water. They may be kept where there is but very little, and only want a pond or tank just deep enough to swim in. The early Aylesbury ducklings that realise such large prices in the London market have hardly ever had a swim; and in rearing ducks, where size is a desideratum, they will grow faster and become larger when kept in pens, farmyards, or in pastures, than where they are at and in the water all day." Where a large number of geese and ducks are kept, water on a sufficient scale, and easily accessible, should be in the neighbourhood.

Ducks, being aquatic birds, do not require heated apartments, nor roosts on which to perch during the night. They squat on the floors, which must be dry and warm. They should, if possible, be kept in a house separate from

Toulouse Goose. Rouen and Aylesbury Ducks.

the other poultry, and it should have a brick floor, so that it can be easily washed. In winter the floor should be littered with a thin layer of straw, rushes, or fern leaves, fresh every day. The hatching-houses should be separated from the lodging apartments, and provided with boxes for the purpose of incubation and hatching.

In its wild state the duck pairs with a single mate: the domestic duck has become polygamous, and five ducks may be allowed to one drake, but not more than two or three ducks should be given to one drake if eggs are required for setting.

Ducks begin laying in January, and usually from that time only during the spring; but those hatched in March will often lay in the autumn, and continue for two or three months. They usually lay fifty or sixty eggs, and have been known to produce 250. The faculty of laying might be greatly developed, as it has been in some breeds of fowls; but they have been hitherto chiefly bred for their flesh. They require constant watching when beginning to lay, for they drop their eggs everywhere but in the nest made for them, but as they generally lay in the night, or early in the morning, when in perfect health, they should therefore be kept in every morning till they have laid. One of the surest signs of indisposition among them is irregularity in laying. "The eggs of the duck," says Mr. Dickson, "are readily known from those of the common fowl by their bluish colour and larger size, the shell being smoother, not so thick, and with much fewer pores. When boiled, the white is never curdy like that of a new-laid hen's egg, but transparent and glassy, while the yolk is much darker in colour. The flavour is by no means so delicate. For omelets, however, as well as for puddings and pastry, duck eggs are much better than hen's eggs, giving a finer colour and flavour, and requiring less butter; qualities so highly esteemed in Picardy, that the women will sometimes go ten or twelve miles for duck eggs to make their holiday cakes."

A hen is often made to hatch ducklings, being considered a better nurse than a duck, which is apt to take them while too young to the pond, dragging them under beetling banks in search of food, and generally leaving half of them in the water unable to get out; and if the fly or the gnat is on the water, she will stay there till after dark, and lose part of her brood. Ducks' eggs may be advantageously placed under a broody exhibition hen. (*See* page 88.) A turkey is much better than either, from the large expanse of the wings in covering the broods, and the greater heat of body; but if the duck is a good sitter, it is best to let her hatch her own eggs, taking care to keep her and them from the water till they are strong. The nest should be on the ground, and in a damp place. Choose the freshest eggs, and place from nine to eleven under her. Feed her morning and evening while sitting, and place food and water within her reach. The duck always covers her eggs upon leaving them, and loose straw should be placed near the house for that purpose.

They are hatched in thirty days. They may generally be left with their mother upon the nest for her own time. When she moves coop her on the short grass if fine weather, or under shelter if otherwise, for a week or ten days, when they may be allowed to swim for half an hour at a time. When hatched they require constant feeding. A little curd, bread-crumbs, and meal, mixed with chopped green food, is the best food when first hatched. Boiled cold oatmeal porridge is the best food for ducklings for the first ten days; afterwards barleymeal, pollard, and oats, with plenty of green food. Never give them hard spring water to drink, but that from a pond. Ducklings are easily reared, soon able to shift for themselves, and to pick up worms, slugs, and insects, and can be cooped together in numbers at night if protected from rats. An old pig-stye is an excellent place for a brood of young ducks.

Ducklings should not be allowed to go on the water till feathers have supplied the place of their early down, for the latter will get saturated with the water while the former throws off the wet. "Though the young ducklings," says Mr. W. C. L. Martin, "take early to the water, it is better that they should gain a little strength before they be allowed to venture into ponds or rivers; a shallow vessel of water filled to the brim and sunk in the ground will suffice for the first week or ten days, and this rule is more especially to be adhered to when they are under the care of a common hen, which cannot follow them into the pond, and the calls of which when there they pay little or no regard to. Rats, weasels, pike, and eels are formidable foes to ducklings: we have known entire broods destroyed by the former, which, having their burrows in a steep bank around a sequestered pond, it was found impossible to extirpate." If the ducklings stay too long in the water they will have diarrhœa, in which case coop them close for a few days, and mix beanmeal or oatmeal with their ordinary food.

A troop of ducks will do good service to a kitchen garden in the summer or autumn, when they can do no mischief by devouring delicate salads and young sprouting vegetables. They will search industriously for snails, slugs, woodlice, and millipedes, and gobble them up eagerly, getting positively fat on slugs and snails. Strawberries, of which they are very fond, must be protected from them. Where steamed food is daily prepared for pigs and cattle, a portion of this mixed with bran and barleymeal is the cheapest mode of satisfying their voracious appetites. They should never be stinted in food.

To fatten ducks let them have as much substantial food as they will eat, bruised oats and peameal being the standard, plenty of exercise, and clean water. Boiled roots mixed with a little barleymeal is excellent food, with a little milk added during fattening. They require neither penning up nor cramming to acquire plumpness, and if well fed should be fit for market in eight or ten weeks. Celery imparts a delicious flavour.

The Aylesbury is the finest breed, and should be of a spotless white, with long, flat, broad beak of a pale flesh colour, grey eyes, long head and neck, broad and flat body and breast, and orange legs, placed wide apart. As it lays early, its ducklings are the earliest ready for market.

They have produced 150 large eggs in a year, and are better sitters than the Rouen.

The Rouen is hardy and easily reared, but rarely lay till February or March. They thrive better in most parts of England than the Aylesburys, and care less for the water than the other varieties. They are very handsome, and weigh eight or nine pounds each, and their flesh is excellent.

The Muscovy duck is so called, says Ray, "not because it comes from Muscovy, but because it exhales a somewhat powerful odour of musk." Little is known of its origin, which is generally thought to be South America; nor has the date of its introduction into Europe been ascertained. "This species," says Mr. W. C. L. Martin, "will interbreed with the common duck, but we believe the progeny are not fertile. The Musk duck greatly exceeds the ordinary kind in size, and moreover, differs in the colours and character of the plumage, in general contour, and the form of the head. The general colour is glossy blue-black, varied more or less with white; the head is crested, and a space of naked scarlet skin, more or less clouded with violet, surrounds the eye, continued from scarlet caruncles on the base of the beak; the top of the head is crested, the feathers of the body are larger, more lax, softer, and less closely compacted together than in the common duck, and seem to indicate less aquatic habits. The male far surpasses the female in size; there are no curled feathers in his tail." The male is fierce and quarrelsome, and when enraged has a savage appearance, and utters deep, hoarse sounds. The flesh is very good, but the breed is inferior as a layer to the Aylesbury or Rouen.

The Buenos Ayres, Labrador, or East Indian, brought most probably from the first-named country, is a small and very beautiful variety, with the plumage of a uniform rich, lustrous, greenish-black, and dark legs and bills; the drake rarely weighing five pounds, and the duck four pounds. Their eggs are often smeared over with a slatey-coloured matter, but the shell is really of a dull white.

CHAPTER XXIII.

GEESE.

Geese require much the same management as ducks. They may be kept profitably where there is a rough pasture or common into which they may be turned, and the pasturage is not rendered bare by sheep, as is generally the case; but even when the pasturage is good, a supply of oats, barley, or other grain should be allowed every morning and evening. Where the pasturage is poor or bad, the old geese become thin and weak, and the young broods never thrive and often die unless fully fed at home. A goose-house for four should not be less than eight feet long by six feet wide and six or seven feet high, with a smooth floor of brick. A little clean straw should be spread over it every other day, after removing that previously used, and washing the floor. Each goose should have a compartment two feet and a half square for laying and sitting, as she will always lay where she deposited her first egg. The house must be well ventilated. All damp must be avoided. A pigstye makes a capital pen. Although a pond is an advantage, they do not require more than a large trough or tank to bathe in.

For breeding not more than four geese should be kept to one gander. Their breeding powers continue to more than twenty years old. It is often difficult to distinguish the sexes, no one sign being infallible except close examination. The goose lays early in a mild spring, or in an ordinary season, if fed high throughout the winter with corn, and on the commencement of the breeding season on boiled barley, malt, fresh grains, and fine pollard mixed up with ale, or other stimulants; by which two broods may be obtained in a year. The common goose lays from nine to seventeen eggs, usually about thirteen, and generally carries straws about previously to laying. Thirteen eggs are quite enough for the largest goose to sit on. They sit from thirty to thirty-five days. March or early April is the best period for hatching, and the geese should therefore begin to sit in February or early March; for goslings hatched at any time after April are difficult to rear. Food and water should be placed near to her, for she sits closely. She ought to leave her nest daily and take a bath in a neighbouring pond. The gander is very attentive, and sits by her, and is vigilant and daring in her defence. When her eggs are placed under a common hen they should be sprinkled with water daily or every other day, for the moisture of the goose's breast is beneficial to them. (*See* page 50.) A turkey is an excellent mother for goslings.

She should be cooped for a few days on a dry grass-plot or meadow, with grain and water by her, of which the goslings will eat; and they should also be supplied with chopped cabbage or beet leaves, or other green food. They must have a dry bed under cover and be protected from rats. Their only dangers are heavy rains, damp floors, and vermin; and they require but little care for the first fortnight; while the old birds are singularly free from maladies of all kinds common to poultry. When a fortnight old they may be allowed to go abroad with their mother and frequent the pond. "It has been formerly recommended," says Mowbray, "to keep the newly-hatched gulls in house during a week, lest they get cramp from the damp earth; but we did not find this indoor confinement necessary; penning the goose and her brood between four hurdles upon a piece of dry grass well sheltered, putting them out late in the morning, or not at all in severe weather, and ever taking them in early in the evening. Sometimes we have pitched double the number of hurdles, for the convenience of two broods, there being no quarrels among this sociable and harmless part of the feathered race. We did not even find it necessary to interpose a parting hurdle, which, on occasion, may be always conveniently done. For the first range a convenient field containing water is to be preferred to an extensive common, over which the gulls or goslings are dragged by the goose, until they become cramped or tired, some of them squatting down and remaining behind at evening." All the hemlock or deadly nightshade within range should be destroyed. When the corn is garnered the young geese may be turned into the stubble which they will thoroughly glean, and many of them will be in fine condition by Michaelmas. Green geese are young geese fattened at about the age of four months, usually on oatmeal and peas, mixed with skim-milk or butter-milk, or upon oats or other grain, and are very delicate. In fattening geese for Christmas give oats mixed with water for the first fortnight, and afterwards barleymeal made into a crumbling porridge. They should be allowed to bathe for a few hours before being killed, for they are then plucked more easily and the feathers are in better condition. Their feathers, down, and quills are very valuable.

Geese are very destructive to all garden and farm crops, as well as young trees, and must therefore be carefully kept out of orchards and plantations. Their dung, though acrid and apt to injure at first, will, when it is mellowed, much enrich the ground.

The Toulouse or Grey Goose is very large, of uniform grey plumage, with long neck, having a kind of dewlap under the throat; the abdominal pouch very much developed, almost touching the ground; short legs; flat feet; short, broad tail; and very upright carriage, almost like a penguin. The Toulouse lays a large number of eggs, sometimes as many as thirty, and even more, but rarely wishes to sit, and is a very bad mother.

The Emden or pure White is very scarce. The bill is flesh-colour, and the legs and feet orange. They require a pond. The Toulouse, crossed with the large white or dark-coloured common breed, produces greater weight than either, and the objection to the former as indifferent sitters and mothers is avoided; but is not desirable for breeding stock, and must have a pond like the White.

CHAPTER XXIV.

DISEASES.

It is more economical to kill at once rather than attempt to cure common fowls showing symptoms of any troublesome disease, and so save trouble, loss of their carcases, and the risk of infection. But if the fowls are favourites, or valuable, it may be desirable to use every means of cure.

See to a sick fowl at once; prompt attention may prevent serious illness, and loss of the bird. When a fowl's plumage is seen to be bristled up and disordered, and its wings hanging or dragging, it should be at once removed from the others, and looked to. Pale and livid combs are as certain a sign of bad health in fowls, as the paleness or lividness of the lips is in human beings. Every large establishment should have a warm, properly ventilated, and well-lighted house, comfortably littered down with clean straw, to be used as a hospital, and every fowl should be removed to it upon showing any symptoms of illness, even if the disease is not infectious, for sick fowls are often pecked at, illtreated, and disliked by their healthy companions. Bear in mind that prevention is better than cure, and that proper management and housing, good feeding, pure water and greens, cleanliness and exercise, will prevent all, or nearly all, these diseases.

Apoplexy arises from over-feeding, and can seldom be treated in time to be of service. The only remedy is bleeding, by opening the large vein under the wing, and pouring cold water on the head for a few minutes. Open the vein with a lancet, or if that is not at hand, with a sharp-pointed penknife; make the incision lengthways, not across, and press the vein with your thumb between the opening and the body, when the blood will flow. If the fowl should recover, feed it on soft, low food for a few days, and keep it quiet. It occurs most often in laying hens, which frequently die on the nest while ejecting the egg; and is frequently caused by too much of very stimulating food, such as hempseed, or improper diet of greaves, and also by giving too much pea or bean meal.

Hard Crop, or being Crop-Bound, is caused by too much food, especially of hard grain, being taken into the crop, so that it cannot be softened by maceration, and is therefore unable to be passed into the stomach. Although the bird has thus too large a supply of food in its crop, the stomach becomes empty, and the fowl eats still more food. Sometimes a fowl swallows a bone that is too large to pass into the stomach, and being kept in the crop forms a kernel, around which fibrous and other hard material collects. Mr. Baily says: "Pour plenty of warm water down the throat, and loosen the food till it is soft. Then give a tablespoonful of castor-oil, or about as much jalap as will lie on a shilling, mixed in butter; make a pill of it, and slide it into the crop. The fowl will be well in the morning. If the crop still remain hard after this, an operation is the only remedy.

The feathers should be picked off the crop in a straight line down the middle. Generally speaking, the crop will be found full of grass or hay, that has formed a ball or some inconveniently-shaped substance. (I once took a piece of carrot three inches long out of a crop.) When the offence has been removed, the crop should be washed out with warm water. It should then be sewn up with coarse thread, and the suture rubbed with grease. Afterwards the outer skin should be served the same. The crop and skin must not be sewed together. For three or four days the patient should have only gruel; no hard food for a fortnight." The slit should be made in the upper part of the crop, and just large enough to admit a blunt instrument, with which you must gently remove the hardened mass.

DIARRHŒA is caused by exposure to much cold and wet, reaction after constipation from having had too little green food, unwholesome food, and dirt. Feed on warm barleymeal, or oatmeal mashed with a little warm ale, and some but not very much green food, and give five grains of powdered chalk, one grain of opium, and one grain of powdered ipecacuanha twice a day till the looseness is checked. Boiled rice, with a little chalk and cayenne pepper mixed, will also check the complaint. When the evacuations are coloured with blood, the diarrhœa has become dysentery, and cure is very doubtful.

GAPES, a frequent yawning or gaping, is caused by worms in the windpipe, which may be removed by introducing a feather, stripped to within an inch of the point, into the windpipe, turning it round quickly, and then drawing it out, when the parasites will be found adhering with slime upon it; but if this be not quickly and skilfully done, and with some knowledge of the anatomy of the parts touched, the bird may be killed instead of cured. Another remedy is to put the fowl into a box, placing in it at the same time a sponge dipped in spirits of turpentine on a hot water plate filled with boiling water, and repeating this for three or four days. Some persons recommend, as a certain cure in a few days, half a teaspoonful of spirits of turpentine mixed with a handful of grain, giving that quantity to two dozen of chickens each day. A pinch of salt put as far back into the mouth as possible is also said to be effectual.

LEG WEAKNESS, shown by the bird resting on the first joint, is generally caused by the size and weight of the body being too great for the strength of the legs; and this being entirely the result of weakness, the remedy is to give strength by tonics and more nourishing food. The quality should be improved, but the quantity must not be increased, as the disease has been caused by over-feeding having produced too much weight for the strength of the legs. Frequent bathing in cold water is very beneficial. This is best effected by tying a towel round the fowl, and suspending it over a pail of water, with the legs only immersed.

LOSS OF FEATHERS is almost always caused by want of green food, or dust-heap for cleansing. Let the fowls have both, and remove them to a grass run if possible. But nothing will restore the feathers till the next moult. Fowls, when too closely housed or not well supplied with green food and lime, sometimes eat each other's feathers, destroying the plumage till the next moult. In such cases green food and mortar rubbish should be supplied, exercise allowed, the injured fowl should be removed to a separate place, and the pecked parts rubbed over with sulphur ointment. Cut or broken feathers should be pulled out at once.

PIP, a dry scale on the tongue, is not a disease, but the symptom of some disease, being only analogous to "a foul tongue" in human beings. Do not scrape the tongue, nor cut off the tip; but cure the roup, diarrhœa, bad digestion, gapes, or whatever the disease may be, and the pip will disappear.

ROUP is caused by exposure to excessive wet or very cold winds. It begins with a slight hoarseness and catching of the breath as if from cold, and terminates in an offensive discharge from the nostrils, froth in the corners of the eyes, and swollen lids. It is very contagious. Separate the fowl from the others, keep it warm, add some "Douglass Mixture" (see "Moulting") to its water daily, wash its head once or twice daily with tepid water, feed it with meal, only mixed with hot ale instead of water, and plenty of green food. Mr Wright advises half a grain of cayenne pepper with half a grain of powdered allspice in a bolus of the meal, or one of Baily's roup pills to be given daily. Mr. Tegetmeier recommends one grain of sulphate of copper daily. Another advises a spoonful of castor-oil at once, and a few hours afterwards one of Baily's roup pills, and to take the scale off the tongue, which can easily be done by holding the beak open with your left hand, and removing the scale with the thumbnail of your right hand; with a pill every morning for a week. If not almost well in a week it will be better to kill it.

THE THRUSH may be cured by washing the tongue and mouth with borax dissolved in tincture of myrrh and water.

PARALYSIS generally affects the legs and renders the fowl unable to move. It is chiefly caused by over-stimulating food. There is no known remedy for this disease, and the fowl seldom if ever recovers. Although chiefly affecting the legs of fowls, it is quite a different disease from LEG WEAKNESS.

VERTIGO results from too great a flow of blood to the head, and is generally caused by over-feeding. Pouring cold water upon the fowl's head, or holding it under a tap for a few minutes, will check this complaint, and the bird should then be purged by a dose of castor-oil or six grains of jalap.

MOULTING.

All birds, but especially old fowls, require more warmth and more nourishing diet during this drain upon their system, and should roost in a warm, sheltered, and properly-ventilated house, free from all draught. Do not let them out early in the morning, if the weather is chilly, but feed them under cover, and give them every morning warm, soft food, such as bread and ale, oatmeal and milk, potatoes mashed up in pot-liquor, with a little pepper and a little boiled meat, as liver, &c., cut small, and a little hempseed with their grain at night. Give them in their water some iron or "Douglass Mixture," which consists of one ounce of sulphate of iron and one drachm of sulphuric acid dissolved in one quart of water; a teaspoonful of the mixture is to be added to each pint of drinking water. This chalybeate is an excellent tonic for weakly young chickens, and young birds that are disposed to outgrow their strength. It increases their appetite, improves the health, imparts strength, brightens the colour of the comb, and increases the stamina of the birds. When chickens droop and seem to suffer as the feathers on the head grow, give them once a day meat minced fine and a little canary-seed.

GROOMBRIDGE'S
SHILLING PRACTICAL MANUALS.

Each Book sent post free for 12 stamps.

---

1. **HOME-MADE WINES.** How to Make and Keep them, with Remarks on preparing the Fruit, fining, bottling, and storing. By G. VINE. Contains Apple, Apricot, Beer, Bilberry, Blackberry, Cherry, Clary, Cowslip, Currant, Damson, Elderberry, Gooseberry, Ginger, Grape, Greengage, Lemon, Malt, Mixed Fruit, Mulberry, Orange, Parsnip, Raspberry, Rhubarb, Raisin, Sloe, Strawberry, Turnip, Vine Leaf, and Mead.

2. **CARVING MADE EASY;** or, Practical Instructions, whereby a Complete and Skilful Knowledge of the Useful Art of Carving may be attained. Illustrated with Engravings of Fish, Flesh, and Fowl, together with Suggestions for the Decoration of the Dinner Table. By A. MERRYTHOUGHT.

3. **COTTAGE COOKERY.** Containing Simple Instructions upon Money, Time, Management of Provisions, Firing, Utensils, Choice of Provisions, Modes of Cooking, Stews, Soups, Broths, Puddings, Pies, Fat, Pastry, Vegetables, Modes of Dressing Meat, Bread, Cakes, Buns, Salting or Curing Meat, Frugality and Cheap Cookery, Charitable Cookery, Cookery for the Sick and Young Children. By ESTHER COPLEY.

4. **COTTAGE FARMING;** or, How to Cultivate from Two to Twenty Acres, including the Management of Cows, Pigs, and Poultry. By MARTIN DOYLE. Contains, On Enclosing a Farm, Land Drainage, Manures, Management of a Two-Acre Farm, Cow Keeping, The Dairy, Pig Keeping, Bees and Poultry, Management of a Ten-Acre Farm, Flax and Rape, Management of a Farm of Twenty Acres, Farm Buildings, etc.

GROOMBRIDGE & SONS, 5, Paternoster Row, London.

GROOMBRIDGE'S
SHILLING PRACTICAL MANUALS.

Each Book sent post free for 12 stamps.

5. **SINGING MADE EASIER FOR AMATEURS,** explaining the pure Italian Method of Producing and Cultivating the Voice; the Management of the Breath; the best way of Improving the Ear; with much other valuable information equally valuable to Professional Singers and Amateurs.

6. **MARKET GARDENING,** giving in detail the various methods adopted by Gardeners in growing the Strawberry, Rhubarb, Filberts, Early Potatoes, Asparagus, Sea Kale, Cabbages, Cauliflowers, Celery, Beans, Peas, Brussels Sprouts, Spinach, Radishes, Lettuce, Onions, Carrots, Turnips, Water Cress, etc. By JAMES CUTHILL, F.R.H.S.

7. **CLERK'S DICTIONARY OF COMMERCIAL TERMS;** containing Explanations of upwards of Three Hundred Terms used in Business and Merchants' Offices. By the Author of "Common Blunders in Speaking and Writing Corrected."
   "An indispensable book for all young men entering a counting-house for the first time."

8. **THE CAT,** its History and Diseases, with Method of Administering Medicine. By the Hon. LADY CUST.

9. **ELOCUTION MADE EASY** for Clergymen, Public Speakers, and Readers, Lecturers, Actors, Theatrical Amateurs, and all who wish to speak well and effectively in Public or Private. By CHARLES HARTLEY. Contents: Cultivation of the Speaking Voice, Management of the Voice, Pausing, Taking Breath, Pitch, Articulation, Pronunciation, The Aspirate, The Letter R, Emphasis, Tone, Movement, Feeling and Passion, Verse, Scriptural Reading, Stammering and Stuttering, Action, Acting, Reciting, etc.

GROOMBRIDGE & SONS, 5, Paternoster Row, London.

GROOMBRIDGE'S
SHILLING PRACTICAL MANUALS.
Each Book sent post free for 12 stamps.

10. **ORATORY MADE EASY,** A Guide to the Composition of Speeches. By CHARLES HARTLEY. Contents: Introduction, Power of Art, Various Kinds of Oratory, Prepared Speech, Constructing a Speech, Short Speeches, Command of Language, Reading and Thinking, Style, Hasty Composition, Forming a Style, Copiousness and Conciseness, Diction or Language, Purity and Propriety, Misapplied Words, Monosyllables, Specific Terms, Variety of Language, Too Great Care about Words, Epithets, Precision, Synonymes, Perspicuity, Long and Short Sentences, Tropes and Figures, Metaphor, Simile, &c.

11. **THE GRAMMATICAL REMEMBRANCER;** or, Aids for Correct Speaking, Writing, and Spelling, for Adults. By CHARLES HARTLEY. Contents: Introduction, Neglect of English Grammar, Divisions of Grammar, Parts of Speech, The Article, The Silent H, Nouns, Formation of the Plural, Genders of Nouns, Cases of Nouns, Comparison of Adjectives, Personal Pronouns, Relative Pronouns, Demonstrative Pronouns, Regular and Irregular Verbs, Shall and Will, The Adverb, Misapplication of Words, Division of Words, Capital Letters, Rules for Spelling Double *l* and *p*, A Short Syntax, Punctuation, &c.

12. **THE CANARY:** Its History, Varieties, Management, and Breeding, with Coloured Frontispiece. By RICHARD AVIS. Contains, History of the Canary, Varieties of the Canary, Food and General Management, Cages, Breeding, Education of the Young, Mules, Diseases, &c.

13. **BIRD PRESERVING** and Bird Mounting, and the Preservation of Birds' Eggs, with a Chapter on Bird Catching. By RICHARD AVIS.

GROOMBRIDGE & SONS, 5, Paternoster Row, London.

Crown 8vo, elegantly bound, gilt edges, Illustrated with 12 beautifully coloured Engravings, price 3s. 6d.

THE CANARY
ITS VARIETIES, MANAGEMENT, AND BREEDING
WITH PORTRAITS OF THE AUTHOR'S OWN BIRDS.
BY THE REV. FRANCIS SMITH.

CONTENTS.

A PLEA FOR THE CANARY	OUR CINNAMONS
ORIGIN OF OUR OWN CANARIA	OUR TURNCRESTS
THE WILD CANARY	THE DOMINIE AND THE GERMANS
OUR LIZARDS	PREPARATIONS FOR BREEDING
OUR YORKSHIRE SPANGLES	NEST BOXES AND NESTS
OUR NORWICH YELLOWS	OUR FIRST BIRDS
OUR LONDON FANCY BIRDS	OUR MISFORTUNES
OUR BELGIANS	OUR INFIRMARY
OUR GREEN BIRDS	ON CAGES

GROOMBRIDGE & SONS, 5, Paternoster Row, London.

Post 8vo, cloth gilt, with Woodcut Illustrations, price 5s.

THE ROSE BOOK
A PRACTICAL TREATISE ON THE CULTURE OF THE ROSE
COMPRISING

The Formation of the Rosarium; the Characters of Species and Varieties; Modes of Propagating, Planting, Pruning, Training, and Preparing for Exhibition; and the Management of Roses in all Seasons.

BY SHIRLEY HIBBERD, F.R.H.S.

CONTENTS.

THE FAMILIES OF WILD ROSES	PILLAR ROSES
THE FAMILIES OF CULTIVATED ROSES	YELLOW ROSES
	ROSES IN POTS
SUMMER ROSES	ROSES IN BEDS
FORMING THE ROSARIUM	ROSES IN GREAT TOWNS
CULTURE OF ROSES IN THE OPEN GROUND	TEA ROSES IN TOWNS
	VARIOUS MODES OF PROPAGATING
AUTUMN PLANTING	SELECT LISTS OF ROSES
SPRING PLANTING	REMINDERS OF MONTHLY WORK IN ROSE GARDEN
PRUNING, DISBUDDING, AND SEASONAL MANAGEMENT	HINTS TO BEGINNERS
CLIMBING ROSES	

GROOMBRIDGE & SONS, 5, Paternoster Row, London.

Crown 8vo, elegantly bound, cloth gilt, Illustrated with 8 beautifully coloured full-page Plates and numerous Wood Engravings, price 3s. 6d.

THE MICROSCOPE

A Popular Description of some of the most Beautiful and Instructive Objects for Exhibition.

With Directions for the Arrangement of the Instruments and the Collection and Mounting of Objects.

BY THE HON. MRS. WARD.

"This elegant book deserves at our hands especial commendation for many reasons. There is no book that we know of that we would more willingly place in the hands of a beginner to create an interest in the science of Microscopy. The Illustrations are beautiful, coloured to represent nature, and all original. To our readers we cannot give better advice than to become purchasers of the book—they will not regret the outlay."—*Electrician.*

GROOMBRIDGE & SONS, 5, Paternoster Row, London.

Crown 8vo, elegantly bound, cloth gilt, Illustrated with 12 beautifully coloured full-page Plates and numerous Wood Engravings, price 3s. 6d.

THE TELESCOPE

A FAMILIAR SKETCH

COMBINING A SPECIAL NOTICE OF OBJECTS COMING WITHIN THE RANGE OF A SMALL TELESCOPE

With a Detail of the most Interesting Discoveries which have been made with the assistance of powerful Telescopes, concerning the Phenomena of the Heavenly Bodies.

BY THE HON. MRS. WARD.

"It is with pleasure that we direct the reader's attention to a little gem lately published by the Hon. Mrs. WARD. One of the most admirable little works on one of the most sublime subjects that has been given to the world. The main design of the book is to show how much may be done in astronomy with ordinary powers and instruments. We have no hesitation in saying that we never saw a work of the kind that is so perfect. The illustrations are admirable, and are all original."—*Western Daily Press.*

GROOMBRIDGE & SONS, 5, Paternoster Row, London.

Crown 8vo, elegantly bound, cloth gilt, Illustrated with 8 beautifully coloured full-page Plates and 90 Wood Engravings, price 3s. 6d.

FIELD FLOWERS

A HANDY BOOK
FOR
THE RAMBLING BOTANIST,
SUGGESTING
WHAT TO LOOK FOR AND WHERE TO GO IN THE OUT-DOOR STUDY OF
BRITISH PLANTS.
BY SHIRLEY HIBBERD, F.R.H.S.

"It will serve as an excellent introduction to the practical study of wild flowers."—*The Queen.*

"We cannot praise too highly the illustrations which crowd the pages of this handbook; the coloured plates are especially attractive, and serve to bring before us very distinctly the most prominent flowers of the field, the heaths, and the hedgerows."—*Examiner.*

GROOMBRIDGE & SONS, 5, Paternoster Row, London.

Crown 8vo, elegantly bound, cloth gilt, illustrated with 8 beautifully coloured Plates and 40 Wood Engravings, price 3s. 6d.

THE FERN GARDEN
HOW TO MAKE, KEEP, AND ENJOY IT
OR,
FERN CULTURE MADE EASY.
BY SHIRLEY HIBBERD, F.R.H.S.

CONTENTS.

FERNS IN GENERAL	MANAGEMENT OF FERN CASES
FERN COLLECTING	THE ART OF MULTIPLYING FERNS
HOW TO FORM AN OUTDOOR FERNERY	BRITISH FERNS
CULTIVATION OF ROCK FERNS	CULTIVATION OF GREENHOUSE STOVE FERNS
CULTIVATION OF MARSH FERNS	SELECT GREENHOUSE FERNS
FERNS IN POTS	SELECT STOVE FERNS
THE FERN HOUSE	TREE FERNS
THE FERNERY AT THE FIRESIDE	FERN ALLIES

GROOMBRIDGE & SONS, 5, Paternoster Row, London.

Crown 8vo, elegantly bound, cloth gilt, Illustrated with 8 full-page coloured Plates and numerous Wood Engravings, price 3s. 6d.

COUNTRY WALKS

OF A NATURALIST

WITH HIS CHILDREN.

By the Rev. W. HOUGHTON, M.A., F.L.S.

"A fresher, pleasanter, or more profitable book than this has rarely issued from the press."—*Art Journal*.

"Contrives to furnish a large amount of interesting natural history in brief compass and in a picturesque and engaging manner."—*Pall Mall Gazette*.

"It is wonderful what a very large amount of most instructive matter connected with the animal and plant world the writer has condensed into a small compass."—*Land and Water*.

"This pretty little volume forms one of the best little books on popular Natural History, and is admirably adapted as a present to the young."—*Birmingham Daily Journal*.

GROOMBRIDGE & SONS, 5, Paternoster Row, London.

Crown 8vo, elegantly bound, cloth gilt, Illustrated with 8 beautifully coloured full-page Plates and numerous Wood Engravings, price 3s. 6d.

SEA-SIDE WALKS

OF A NATURALIST

WITH HIS CHILDREN.

By the Rev. W. HOUGHTON, M.A., F.L.S.

"The wonders of the sea-shore are detailed in an easy, pleasant, and lucid style."—*Examiner*.

"The book is very attractive, and its usefulness is enhanced by its many careful illustrations."—*Daily Telegraph*.

"Families visiting the sea-side should provide themselves with this convenient and instructive work."—*The Queen*.

"It is pleasingly written, and the scientific information is correct and well selected."—*Athenæum*.

GROOMBRIDGE & SONS, 5, Paternoster Row, London.

BOOKS FOR YOUNG NATURALISTS.

Crown 8vo, elegantly bound, gilt edges, Illustrated with 16 beautifully coloured Plates and numerous Wood Engravings, price 5s.

NESTS AND EGGS
OF FAMILIAR BIRDS.

Described and Illustrated with an account of the Haunts and Habits of the Feathered Architects, and their Times and Modes of Building.

By H. G. ADAMS.

GROOMBRIDGE & SONS, 5, Paternoster Row, London.

Crown 8vo, elegantly bound, gilt edges, Illustrated with 8 beautifully coloured Plates and numerous Wood Engravings, price 3s. 6d.

BEAUTIFUL BUTTERFLIES.
DESCRIBED AND ILLUSTRATED

With an Introductory chapter, containing the History of a Butterfly through all its Changes and Transformations. A Description of its Structure in the Larva, Pupa, and Imago states, with an Explanation of the scientific terms used by Naturalists in reference thereto, with observations upon the Poetical and other associations of the Insect.

By H. G. ADAMS.

GROOMBRIDGE & SONS, 5, Paternoster Row, London.

Crown 8vo, elegantly bound, gilt edges, Illustrated with 8 beautifully coloured Plates and numerous Wood Engravings, price 3s. 6d.

BEAUTIFUL SHELLS
THEIR NATURE, STRUCTURE, AND USES FAMILIARLY EXPLAINED.

With Directions for Collecting, Clearing and Arranging them in the Cabinet.

Descriptions of the most remarkable Species, and of the creatures which inhabit them, and explanations of the meaning of their scientific names, and of the terms used in Conchology.

By H. G. ADAMS.

GROOMBRIDGE AND SONS, 5, Paternoster Row, London.

Crown 8vo, elegantly bound, gilt edges, Illustrated with 8 beautifully coloured Plates and Wood Engravings, price 3s. 6d.

HUMMING BIRDS.
DESCRIBED AND ILLUSTRATED.
WITH AN

Introductory Sketch of their Structure, Plumage, Faunts, Habits, etc.

By H. G. ADAMS.

GROOMBRIDGE & SONS, 5, Paternoster Row, London.

PRICE ONE SHILLING (post free for Thirteen Stamps).

PIPER'S POULTRY-YARD ACCOUNT BOOK.

A SIMPLE PLAN FOR KEEPING A CORRECT ACCOUNT OF

EXPENDITURE AND RECEIPTS,

ALSO FOR SHOWING WHAT EACH ITEM AMOUNTS TO IN THE WHOLE YEAR.

ADAPTED FOR ANY YEAR, AND FOR BEGINNING AT ANY TIME OF THE YEAR.

By HUGH PIPER,

Author of 'Poultry, a Practical Guide,' and 'Pigeons.'

With the Aid of this Account Book, the following Statistics will be able to be determined:—

The Number of Eggs Laid Daily—The Total Amount Received for Produce—The Number of Eggs Sold—The Amount Received for Eggs—The Number of Fowls Sold—The Amount Received for Fowls—The Number of Chickens Sold—The Amount Received for Chickens—The Value of Feathers and Manure—The Number of Eggs Used in the Household—The Number of Fowls Used in the Household—The Total Amount of Expenditure—The Number of Fowls Purchased—The Value of Eggs Purchased for Setting—The Cost of Food, Rent, Labour, and other Sundries—The Number of Hens Set—The Different Dates of Setting—The Number of Eggs—The Dates when Due—The Number of Chickens Hatched, and the Number of Chickens Reared.

The Balance Sheet at the end determining whether the transactions of the Yard have been carried out at a profit or loss.

GROOMBRIDGE AND SONS, 5, Paternoster Row, London.

Date_____ 18____

To Mr._____

       Bookseller,

_____

Please to send me _____ copies of 'PIPER'S POULTRY-YARD ACCOUNT BOOK.' I enclose_____ in payment for the same.

    Name_____

    Address_____

*** This Order can be filled up and forwarded to any Bookseller; or to the Publishers, GROOMBRIDGE & SONS, 5, Paternoster Row, London.

PUBLISHERS' NOTICE.

Subscribers who approve of this Magazine will much assist the Editor and Publishers by forwarding this Notice with a recommendation to any friend interested in Gardening.

WITH COLOURED PLATES.
Valuable Work of Reference for the Garden and Greenhouse.

THE FLORAL WORLD
AND
GARDEN GUIDE.
EDITED BY
SHIRLEY HIBBERD, Esq., F.R.H.S.

PUBLISHED MONTHLY, PRICE SIXPENCE.

A specimen number sent post free for Seven Stamps.

THE FLORAL WORLD is devoted entirely to Gardening Subjects, as represented in the several departments of Plant Houses, Flower, Fruit, and Vegetable Culture, Garden Scenes and Embellishments, the Management of Allotment Lands, Flower Shows, and Horticultural Botany. These are severally treated in a simple and practical manner by experienced pens, and the fullest attention is given to communications from Correspondents, whether seeking or conveying information.

Each Number contains a highly finished Coloured Plate, besides Woodcut Illustrations.

THE ANNUAL SUBSCRIPTION is SIX SHILLINGS,

Which sum can either be paid to any Bookseller, or can be forwarded direct to the Publishers, when the 12 Numbers for the year will be sent regularly (post paid) as issued. THE FLORAL WORLD is published on the 1st of each month, and is the only Gardening Magazine at the price containing Coloured Plates.

GROOMBRIDGE & SONS, 5, Paternoster Row, London, and all Booksellers.

For the convenience of intending Subscribers the following FORM OF ORDER *is appended :—*

Date_____18__

To Mr._____
    *Bookseller,*_____

Please to send me 'THE FLORAL WORLD AND GARDEN GUIDE' regularly monthly as issued, for Twelve Months, commencing with the_____number.

I enclose 6s. in payment for the same.

Name_____

Address_____

** This Form of Order can be filled up and forwarded to any Bookseller; or to the Publishers, GROOMBRIDGE & SONS, 5, Paternoster Row, London.

www.ingramcontent.com/pod-product-compliance
Lightning Source LLC
Chambersburg PA
CBHW020246170426
43202CB00008B/250